# Remote Pilot Career Guide:

From Beginner to Professional Drone Captain Step By Step Guide

[Professional Edition 2026]

Capt Shekhar Gupta

Airplane and Remote Pilot

*Training,* • *Certification,* • *Aviation, Knowledge,* • *Drone Operations,* • *Jobs,* • *Business,* • *Global Opportunities,*



notionpress.com

INDIA • SINGAPORE • MALAYSIA

ISBN
Paperback 979-8-90584-241-2
Hardcase 979-8-90584-242-9

# Contents


| | | |
|---|---|---|
| *Author Bio* | | *7* |
| *Preface* | | *11* |
| **PART I: INTRODUCTION TO THE DRONE WORLD** | | |
| Chapter 1: | Introduction to Drones | 19 |
| Chapter 2: | Drone Industry Overview | 27 |
| Chapter 2A: | Drone Industry Overview | 34 |
| Chapter 3: | Understanding Drone Terminology | 43 |
| **PART II: DRONE TYPES & TECHNOLOGY** | | |
| Chapter 4: | Anatomy of a Drone | 55 |
| Chapter 5: | Types of Drones by Design | 63 |
| Chapter 6: | Types of Drones by Weight Category | 75 |
| Chapter 7: | Types of Drones by Purpose | 83 |
| Chapter 8: | Leading Drone Manufacturers | 91 |
| **PART III: DRONE REGULATIONS & LEGAL FRAMEWORK** | | |
| Chapter 9: | Drone Rules in India | 99 |
| Chapter 9A: | Drone Rules in India | 105 |
| Chapter 10: | Understanding Airspace | 111 |
| Chapter 11: | Drone Registration & Compliance | 117 |
| Chapter 12: | Privacy, Ethics & Data Protection | 123 |
| **PART IV: BECOMING A CERTIFIED DRONE PILOT** | | |
| Chapter 13: | Becoming a Certified Drone Pilot | 133 |
| Chapter 14: | DGCA Remote Pilot Training Organizations [RPTOs] | 139 |

Chapter 15: DGCA Drone Pilot Certification Process 145

Chapter 16: Drone Pilot License & Career Roadmap 151

**PART V: DRONE PILOT**

Chapter 17: Air Regulations 161

Chapter 18: Principles of Flight 167

Chapter 19: Airspace & Navigation 173

Chapter 20: Aviation Meteorology Guide for Drone Pilots 179

Chapter 21: Radio Telephony & Communication 185

Chapter 22: Drone Maintenance 193

Chapter 23: Simulator Training 201

Chapter 24: Practical Flight Training 207

Chapter 25: Risk Assessment & Safety Management 213

**PART VI: AVIATION FUNDAMENTALS FOR DRONE PILOTS**

Chapter 26: Introduction to Aviation 223

Chapter 26A: Introduction to Aviation 229

Chapter 27: Air Traffic Control Basics 235

Chapter 28: Aviation Safety Culture 241

Chapter 29: Flight Planning 247

**PART VII: OPERATIONS & ADVANCED FLYING**

Chapter 30: Pre-Flight Procedures 255

Chapter 31: Standard Flight Operations — The Professional Drone Pilot's Field Guide 263

Chapter 32: Autonomous Flight Operations 271

Chapter 33: BVLOS Operations 277

Chapter 34: Night Operations 285

Chapter 35: Emergency Operations 291

*Disclaimer* *297*

The Drone revolution is transforming industries worldwide. From agriculture and infrastructure inspection to logistics, filmmaking, disaster management, and smart cities, the demand for skilled Remote Pilots has never been greater.

Written by experienced Aviation professional Capt. Shekhar Gupta, this comprehensive guide provides aspiring Drone Pilots, Aviation students, entrepreneurs, and industry professionals with a practical roadmap to succeed in the rapidly evolving UAV ecosystem.

Inside this book, you will discover:

✓ Understanding Drone Technology and UAV Systems

✓ DGCA Regulations and Drone Rules in India

✓ Remote Pilot Certification Pathways

✓ Career Opportunities Across Multiple Industries

✓ Drone Mapping, Surveying, and Inspection Applications

✓ AI-Powered Drone Technologies

✓ Starting and Scaling a Drone Business

✓ International Opportunities for Drone Professionals

✓ Future Trends Shaping Aviation and Autonomous Flight

Whether you are a student exploring Aviation careers, a Pilot seeking new opportunities, or an entrepreneur looking to enter the Drone industry, this handbook offers the knowledge, strategies, and insights needed to build a successful future.

# Author Bio

**Capt. Shekhar Gupta**

**Airplane and Remote Pilot | Aviation Author | Drone Industry Mentor**

**Capt. Shekhar Gupta** is an accomplished Aviation professional, entrepreneur, and prolific author with extensive experience spanning both manned Aircraft operations and uncrewed Aerial vehicle (UAV) systems. As a professional Pilot with over eight years of experience, he has operated fourteen different types of Aircraft across ten countries, successfully bridging the disciplined world of traditional civil Aviation with emerging remote Flight Technologies.

Capt. Shekhar serves as the CEO and Founder of a unified Aviation, Technology, and transport ecosystem, leading ventures such as **Asiatic International Corp**, **AeroSoft Corp**, and **Air-Aviator.com**.

A passionate educator, career counselor, and personal branding mentor, he has dedicated his career to building operational excellence, developing remote-first business models, and structuring cutting-edge training programs ranging from cabin crew preparation to DGCA-approved drone Pilot certifications.

An established industry voice, Capt. Shekhar has authored and co-authored an extensive catalog of 15+ published Books specializing in Aviation pathways, Start-Up leadership, Human Resources, and Business Strategy. His published works are widely distributed across global platforms including Amazon and through his dedicated publishing network at AlfaBook.AsiaticInCorp.org. Through this comprehensive guide, he synthesizes years of practical insights, complex regulatory frameworks, and market trends to provide an actionable roadmap for the next generation of aerospace pioneers.

## Published Books Bibliography (15-Book Catalog)

### ✈ Aviation & Cabin Crew Career Guides

1. **Pilot's Career Guide**

   The definitive resource on international Aviation training, financing strategies, and corporate or Air charter employment opportunities. Includes a comprehensive FAQ section covering common Pilot interview questions and critical mistakes to avoid.

2. **Learn to Fly Here: Your Pilot Career Guide**

   A complete, simplified roadmap designed to turn Pilot dreams into reality. This guide demystifies the path to becoming a commercial Pilot, walking candidates from their first trial Flight straight to the Airline cockpit.

3. **Flying High: A Guide to a Career in Aviation Industry**

   An insightful handbook that propels aspiring professionals into the commercial Airline landscape, offering practical guidance for navigating various operational roles within dynamic global Airlines.

4. **International Air Hostess Career Guide**

   Co-authored with seasoned ex-cabin crew member Sabita Lalung, this serves as an ultimate passport to a sky-high career, teaching hospitality excellence, safety standards, and confidence for world-class cabin crew roles.

### 🏢 Startup Leadership, HR, & Corporate Strategy

5. **From Dreamer To CEO: A Step-by-Step Roadmap to Leadership, Strategy, and Startup Success**

   A practical, zero-theory guide drawing from Capt. Gupta's experience managing 16 remote, dot-com startups. It outlines actionable steps required to transition a visionary idea into an operational business.

6. **Building HR Excellence: A Startup Founder's Guide**

   A no-nonsense lifeline for founders to build robust human resource infrastructure from day one, mitigating high turnover, protecting company culture, and managing startup legal risks.

7. **Lessons for Business Leaders**

   A concise and engaging collection of essential strategies and wisdom forged in today's competitive landscape, sharing vital takeaways on agile leadership and corporate endurance.

🎓 **Professional Mentorship & Multidisciplinary Growth**

8. **All Best Career Guide** A continuous-evaluation handbook that helps students map their inherent strengths and capabilities to industry demands, answering daily career queries across various training sectors.

9. **Global Educator Series**

   A specialized series tracking international trends in educational design, instructional Technology, and modern teaching methodologies across global boundaries.

🌟 **Leadership, Diversity, & Entrepreneurship Showcases**

10. **Asia's Most Savvy Women (2nd Edition)** A powerful, inspiring celebration of female entrepreneurs, researchers, healers, and changemakers across Asia, exploring the unique mindsets and tactical decisions behind their cross-industry success.

11. **Most Successful Female Entrepreneurs of India** A curated compilation of life lessons from Indian women leading sectors from the army to Aviation, astrology to accountancy, and blogging to professional car racing.

*Note: For the complete 15+ title extended catalog, multi-volume textbook sets, and regional tourism directories (including the Pan-India Tourism Network Directory), please visit the official digital archive at AlfaBook. AsiaticInCorp.org*

## Title & Publication Metadata

**REMOTE Pilot CAREER GUIDE 2026**

*Build Your Future in the Drone Industry*

**Core Disciplines Covered:**

Training • Certification • Careers • Business • AI • Future Aviation

**Capt. Shekhar Gupta**

*Airplane and Remote Pilot*

**Published by:**

AeroSoft Corp

Headquarters: Leeds Enclave, Airport Road, Indore, Madhya Pradesh, India

https://alfabook.asiaticincorp.org

---

**Remote Pilot Career Guide 2026**

**Build Your Future in the Drone Industry**

**Training • Certification • Careers • Business • AI • Future Aviation**

Capt. Shekhar Gupta

*Airplane and Remote Pilot*

# Preface

Remote Pilot Career Guide 2026

The Aviation industry is experiencing one of the most significant transformations in its history. For more than a century, Flight was associated primarily with manned Aircraft operated by trained Pilots. Today, a new era has emerged—an era in which Remotely Piloted Aircraft systems, commonly known as Drones, are reshaping industries, creating new career opportunities, and redefining the future of Aviation.

As an Airplane and Remote Pilot, I have witnessed the evolution of Aviation Technology and the increasing integration of Drones into commercial, industrial, governmental, and humanitarian operations. What was once considered a niche Technology has become a vital tool for agriculture, infrastructure inspection, surveying, logistics, media production, disaster management, environmental conservation, and countless other applications.

This book was written to provide Students, Professionals, Entrepreneurs, and Aviation enthusiasts with a practical Roadmap for understanding and succeeding in the rapidly growing Indian and Global Drone Ecosystem.

The coming decade will witness unprecedented growth in unmanned Aviation Organizations across the globe are investing heavily in Drone Technology, artificial intelligence, autonomous navigation systems, and Aerial data analytics. The demand for skilled Remote Pilots and Drone professionals is expected to rise dramatically.

The objective of this book is not merely to explain Drone operations but to help readers build meaningful careers within this exciting industry.

Whether you are a student exploring future career options, a Pilot seeking new opportunities, an Entrepreneur planning a Drone venture, or a Professional interested in emerging Technologies, this guide aims to provide valuable insights, practical knowledge, and actionable strategies.

I invite you to embark on this journey into the future of Aviation.

**Capt. Shekhar Gupta**

Airplane and Remote Pilot

The Global Drone industry has transitioned from experimentation to mainstream adoption. Governments, businesses, Educational Institutions, and individuals increasingly recognize the potential of Drone Technology to solve complex challenges and improve operational efficiency.

In India, supportive regulations, government initiatives, Technological advancements, and growing commercial demand have created a favorable environment for Drone innovation and career development.

Remote Pilots now have opportunities in:

- Precision Agriculture
- Surveying and Mapping
- Construction Monitoring
- Mining Operations
- Energy Infrastructure Inspection
- Telecommunications
- Environmental Monitoring
- Emergency Response
- Law Enforcement
- Media and Entertainment
- Logistics and Delivery Services

The profession of a Remote Pilot combines Aviation knowledge, Technological expertise, operational discipline, and problem-solving skills.

Unlike many traditional professions, the Drone industry offers multiple career pathways:

- Employment
- Freelancing
- Consulting
- Training
- Entrepreneurship
- Technology Development

This guide explores each of these pathways in detail while preparing readers for the future of autonomous Aviation.

# PART I

# INTRODUCTION TO THE DRONE WORLD

# CHAPTER 1: INTRODUCTION TO DRONES

- **What is a Drone?**
- **What is UAS, UAV & RPAS?**
- **Evolution of Drone Technology**
- **History of Drones**
- **Global Drone Revolution**
- **Drone Industry in India**

# Chapter 1 Introduction to Drones

## What is a Drone?

At its simplest, a Drone is any vehicle that can travel through its environment without a human operator physically on board. In the context of modern Aviation, the word refers to an uncrewed Aircraft capable of Flight either through Remote manual manipulation by a Pilot on the ground, or completely autonomously via pre-programmed Flight software and integrated onboard sensors.

While early iterations of this Technology resembled simple radio-controlled model Airplanes, modern Drones are essentially highly sophisticated Flying computers. They are equipped with global positioning systems, advanced inertial measurement units, microprocessors, and payload bays capable of lifting complex industrial equipment. The defining trait of a true Drone is its ability to maintain stable, controlled Flight and execute complex commands without direct, continuous structural feedback from a physical cockpit.

## Understanding the Acronyms:

### UAS, UAV, and RPAS

To build a professional career at the executive level of this industry, you must move past everyday consumer terms and master the precise Technical and legal language of global aerospace. Civil Aviation authorities, corporate clients, and insurance syndicates do not use the word Drone in official operational contracts or regulatory filings. They rely on three distinct acronyms, each carrying specific legal and structural definitions.

The first acronym is UAV, which stands for Unmanned Aerial Vehicle. This term refers strictly to the physical Airframe that leaves the ground. It encompasses the carbon-fiber chassis, the electric brushless motors or fuel engines, the propellers, the onboard electronic speed controllers, and the primary internal Flight management unit. A UAV is simply the piece of hardware; on its own, it cannot execute a mission.

The second acronym is UAS, which stands for Unmanned Aircraft System. This is the overarching term preferred by major regulatory bodies like the Directorate General of Civil Aviation in India and the Federal Aviation Administration in the United States. A UAS recognizes that safe Flight requires an entire supporting infrastructure. Therefore, a UAS includes the physical UAV, the Ground Control Station consisting of the Remote controller, telemetry screen, or laptop, the dedicated wireless data links that transmit commands and live video signals, and the operational payload such as a thermal imaging camera or agricultural sprayer.

The third acronym is RPAS, which stands for Remotely Piloted Aircraft System. This terminology is utilized by the International Civil Aviation Organization to highlight a crucial legal and operational reality: that a licensed human Pilot is actively in command of the Aircraft at all times, even if the system is executing an automated Flight path. By defining a system as an RPAS, international law integrates the Drone directly into civil Airspace frameworks, placing ultimate legal and operational accountability squarely on your shoulders as the Remote Pilot in Command.

## The Evolution of Drone Technology and a History of the Skies

The journey of uncrewed Aviation is a narrative of Technological convergence. Drones did not emerge suddenly; they are the byproduct of over a century of military experimentation, radio-controlled hobbyist innovations, and the mass-market smartphone computing revolution.

The earliest ancestral roots of unmanned Flight trace back to the First and Second World Wars. In 1917, the United States military developed the Kettering Bug, an uncrewed Flying bomb guided by internal pneumatic gyroscopes and preset mechanical counters. By the 1930s, the British Royal Navy built the DH.82B Queen Bee, a radio-controlled biplane utilized as a live target for anti-Aircraft gunners. The constant, low buzzing sound of the Queen Bee's engine over the water led engineers to coin the term Drone, a moniker that would stay with the Technology for decades.

For the remainder of the twentieth century, uncrewed Flight was strictly divided between highly classified, multimillion-dollar military reconnaissance operations and passionate radio-control hobbyists assembling fragile balsa-wood kits. The massive pivot occurred in the late 2000s and early 2010s, driven entirely by the smartphone manufacturing explosion. As global Tech giants spent billions of dollars to make smartphone components smaller, lighter, and vastly more powerful, the core electronics required for stable Drone Flight suddenly became cheap and mass-produced.

Micro-gyroscopes, compact GPS sensors, light lithium-polymer batteries, and miniature digital cameras were suddenly available at a fraction of their historical cost. Early Drone manufacturers realized that by clustering these tiny sensors inside an automated multi-rotor frame, they could create an incredibly stable, intelligent Aerial camera platform that required zero assembly and minimal training to hover. Over the next decade, the Technology matured rapidly, shifting from a simple novelty for hobbyists into an indispensable industrial data gathering asset.

## The Global Drone Revolution

We are living through a massive global shift in how Aerial data is acquired and how logistics are handled. Drones have permanently shattered the traditional barriers to the sky. Historically, if a commercial enterprise required an Aerial survey of a mining site, a high-resolution map of an infrastructure project, or a thermal sweep of a massive agricultural estate, they had to lease a full-sized helicopter or Airplane, secure an Airport slot, and hire an entire Flight crew. The cost was astronomical, and the logistical friction was immense.

Today, a single certified Remote Pilot in Command can pull up to a site, deploy an enterprise-grade Unmanned Aircraft System from the back of a utility vehicle, and execute a high-precision digital survey within an

Hour at a fraction of the historical cost. This drastic reduction in cost and complexity has sparked an absolute commercial boom.

Globally, the commercial Drone sector is growing at an incredible pace, driven by massive investments in automation, logistics, and clean Technology. Drones are no longer just passive cameras; they are active tools. In modern logistics, automated cargo Drones are completing long-distance beyond visual line of sight deliveries, carrying medical supplies across rural regions, and transforming supply chain efficiency. In industrial asset management, specialized Drones equipped with artificial intelligence are replacing dangerous manual inspections, mapping vast open-cast mines, identifying structural micro-fractures in wind turbines, and monitoring pipeline corridors around the globe.

## The Drone Industry in India

The scale of the Drone ecosystem in India reflects a deliberate, structural transformation driven by progressive policy reforms, robust domestic manufacturing incentives, and wide-scale public and private sector adoption. The baseline foundation for this boom was established through the liberalized Drone Rules, which significantly lowered entry barriers for professional operators by replacing complex, multi-layered clearance procedures with a streamlined, digital interface.

The official government data highlights the explosive growth of this regulated ecosystem. India has built an expansive Drone network featuring over thirty-eight thousand registered Drones holding official Unique Identification Numbers, supported by an active workforce of nearly forty thousand DGCA-certified Remote Pilots, and trained across more than two hundred forty approved Remote Pilot Training Organisations nationwide.

This rapid expansion is further accelerated by powerful economic and manufacturing policies. The Government of India established a dedicated Production Linked Incentive scheme to aggressively scale domestic manufacturing of Drones and Drone components, while simplifying corporate tax landscapes by lowering the uniform GST on Drones to five percent. This structural framework aims to position India as a premier global Drone hub by the end of the decade.

The commercial application of this Technology across India is visible across three major pillars. The first is public sector transformation, highlighted by the nationwide SVAMITVA scheme, which utilizes high-resolution mapping Drones to survey over three lakh villages, creating accurate property cards for millions of rural citizens. The second is the monetization of agricultural services, underscored by the Namo Drone Didi initiative, which deploys thousands of advanced spraying Drones to women-led self-help groups, driving modern precision agriculture across rural communities. The third is the massive expansion of the Drone-as-a-Service model across industrial logistics, highway development mapping, and infrastructure monitoring, establishing the Drone industry as a cornerstone of India's digital engineering future.

# CHAPTER 2: DRONE INDUSTRY OVERVIEW

- **Global Market Size**
- **Indian Drone Ecosystem**
- **Government Initiatives**
- **Drone Startups**
- **Emerging Trends**
- **Future Growth Projections**

# Chapter 2 Drone Industry Overview

## Global Market Size and Economic Impact

The unmanned Aviation sector has officially transitioned from an emerging niche into a driving pillar of the global aerospace economy. The total global Drone market—spanning military, commercial, and consumer hardware, software, and services—is valued at approximately 100.74 billion dollars. Driven by rapid automation, regulatory ease, and deep integration into industrial workflows, the market is on a trajectory to expand to over 210 billion dollars by the early 2030s, sustaining a steady compound annual growth rate [CAGR] of nearly 9.6%.

Within this overarching footprint, the commercial Drone segment is experiencing the fastest rate of capital injection. Valued at 28 billion dollars, the commercial sector is expanding at an explosive annual rate of more than 9.3%. Regionally, the Asia-Pacific territory has emerged as the dominant powerhouse, capturing over 31.6% of global revenue. This leadership is fueled by intense industrial digitization, massive infrastructure projects, and large-scale agricultural adoption across developing economies.

From an operational standpoint, hardware continues to command the largest share of market value at roughly 61%, led by the mass manufacturing of rotary-blade systems which hold 78% of the market due to their unique hovering capabilities. However, the true high-margin frontier is shifting rapidly toward Drone-as-a-Service [DaaS] business models, cloud-based telemetry analytics, and hybrid VTOL platforms, which grew by more than 22% over the past year alone.

## The Indian Drone Ecosystem

India represents one of the most dynamic and aggressively growing Drone ecosystems in the world. Once constrained by tightly restrictive import laws and manual multi-agency approval structures, the domestic market has completely opened up. Today, India's commercial Drone sector alone is valued at 1.2 billion dollars, with projections positioning it to double to 2.4 billion dollars over the next several years, reflecting a robust growth trajectory. When factoring in the heavy modernization of defense and border reconnaissance platforms, the total Indian UAV market is expected to cross 1.8 billion dollars.

The architecture of India's ecosystem is unique because it is heavily driven by a standardized, digital-first infrastructure. Rather than scaling via fragmented, unregulated private networks, the Indian Drone industry is anchored firmly around the eGCA and Digital Sky frameworks. This structural centralization has turned the country into a fertile laboratory for automated Technology. The domestic ecosystem now features over 38,000 registered Drones holding valid Unique Identification Numbers, an active operational workforce of nearly 40,000 certified Remote Pilots, and a rapidly expanding training foundation consisting of more than 240 DGCA-approved Remote Pilot Training Organisations nationwide.

## Transformative Government Initiatives

The rapid scaling of India's Drone sector is the direct result of deliberate, coordinated legislative policies designed to eliminate operational friction and foster self-reliance. The transition began with the liberalization of the Drone Rules, which replaced over 90% of traditional compliance forms with a trust-based electronic filing system and reduced uniform processing fees to nominal levels. This baseline was further strengthened by a flat rationalization of the Goods and Services Tax [GST] to 5% for all certified Drone systems.

To catalyze domestic manufacturing and eliminate reliance on foreign supply chains, the Ministry of Civil Aviation instituted the Production Linked Incentive [PLI] scheme for Drones and Drone components. This policy provides direct financial incentives to local manufacturers, sparking massive domestic research into localized Flight controllers, propulsion motors, and carbon-fiber composite manufacturing. This is closely pAired with a strict import ban on foreign-assembled conventional

Drones, ensuring that public and private capital flows directly into Indian manufacturing pipelines.

On the application front, two massive public programs are driving nationwide adoption:

- **The SVAMITVA Scheme:** A monumental land-mapping project executed by the Ministry of Panchayati Raj. Utilizing high-precision mapping Drones equipped with RTK navigation, the initiative has successfully surveyed over three lakh villages across India, creating highly accurate digital property cards and providing secure asset ownership documentation to millions of rural citizens.

## Global Market Size and Economic Impact

The unmanned Aviation sector has officially transitioned from an emerging niche into a driving pillar of the global aerospace economy. The total global Drone market—spanning military, commercial, and consumer hardware, software, and services—is valued at approximately 100.74 billion dollars. Driven by rapid automation, regulatory ease, and deep integration into industrial workflows, the market is on a trajectory to expand to over 210 billion dollars by the early 2030s, sustaining a steady compound annual growth rate [CAGR] of nearly 9.6%.

Within this overarching footprint, the commercial Drone segment is experiencing the fastest rate of capital injection. Valued at 28 billion dollars, the commercial sector is expanding at an explosive annual rate of more than 9.3%. Regionally, the Asia-Pacific territory has emerged as the dominant powerhouse, capturing over 31.6% of global revenue. This leadership is fueled by intense industrial digitization, massive infrastructure projects, and large-scale agricultural adoption across developing economies.

From an operational standpoint, hardware continues to command the largest share of market value at roughly 61%, led by the mass manufacturing of rotary-blade systems which hold 78% of the market due to their unique hovering capabilities. However, the true high-margin frontier is shifting rapidly toward Drone-as-a-Service [DaaS] business models, cloud-based telemetry analytics, and hybrid VTOL platforms, which grew by more than 22% over the past year alone.

## The Indian Drone Ecosystem

India represents one of the most dynamic and aggressively growing Drone ecosystems in the world. Once constrained by tightly restrictive import laws and manual multi-agency approval structures, the domestic market has completely opened up. Today, India's commercial Drone sector alone is valued at 1.2 billion dollars, with projections positioning it to double to 2.4 billion dollars over the next several years, reflecting a robust growth trajectory. When factoring in the heavy modernization of defense and border reconnaissance platforms, the total Indian UAV market is expected to cross 1.8 billion dollars.

The architecture of India's ecosystem is unique because it is heavily driven by a standardized, digital-first infrastructure. Rather than scaling via fragmented, unregulated private networks, the Indian Drone industry is anchored firmly around the eGCA and Digital Sky frameworks. This structural centralization has turned the country into a fertile laboratory for automated Technology. The domestic ecosystem now features over 35,000+ registered Drones holding valid Unique Identification Numbers, an active operational workforce of nearly 50,000+ certified Remote Pilots, and a rapidly expanding training foundation consisting of more than 230+ DGCA-approved Remote Pilot Training Organisations [RTPO] nationwide.

## Transformative Government Initiatives

The rapid scaling of India's Drone sector is the direct result of deliberate, coordinated legislative policies designed to eliminate operational friction and foster self-reliance. The transition began with the liberalization of the Drone Rules, which replaced over 90% of traditional compliance forms with a trust-based electronic filing system and reduced uniform processing fees to nominal levels. This baseline was further strengthened by a flat rationalization of the Goods and Services Tax [GST] to 5% for all certified Drone systems.

To catalyze domestic manufacturing and eliminate reliance on foreign supply chains, the Ministry of Civil Aviation instituted the Production Linked Incentive [PLI] scheme for Drones and Drone components. This policy provides direct financial incentives to local manufacturers, sparking massive domestic research into localized Flight controllers, propulsion motors, and carbon-fiber composite manufacturing. This is closely pAired with a strict import ban on foreign-assembled conventional Drones, ensuring that public and private capital flows directly into Indian manufacturing pipelines.

On the application front, two massive public programs are driving nationwide adoption:

- **The SVAMITVA Scheme:** A monumental land-mapping project executed by the Ministry of Panchayati Raj. Utilizing high-precision mapping Drones equipped with RTK navigation, the initiative has successfully surveyed over three lakh villages across India, creating highly accurate digital property cards and providing secure asset ownership documentation to millions of rural citizens.

Pioneering firms like IdeaForge Technology have established immense market dominance, scaling from early-stage academic spin-offs into major defense and enterprise suppliers capable of listing publicly on national stock exchanges. Alongside them, companies like Asteria Aerospace, Aarav Unmanned Systems [AUS], and Garuda Aerospace have successfully commercialized enterprise-grade mapping and agricultural software solutions.

Concurrently, deep-Tech startups like Detect Technologies have integrated industrial Drone workflows with automated machine learning pipelines, allowing automated Drones to spot microscopic structural defects, pipeline leaks, and corrosion across massive oil refineries and power plants without human intervention. This robust startup environment has attracted significant venture capital, with early-stage funding increasingly focusing on specialized payload development, long-endurance propulsion systems, and Drone logistics infrastructure.

## Emerging Technological Trends

The state of uncrewed Aviation is being continuously reshaped by the convergence of automation, edge computing, and alternative energy sources. As a professional Drone Captain, keeping pace with these emerging trends is essential for maintaining a competitive edge in the commercial job market.

- **Artificial Intelligence and Autonomous Navigation:** Modern Drones are moving away from continuous manual radio guidance. Integrated with Edge AI processors and advanced computer vision, 2026-generation Drones can navigate safely through dense, complex environments even when completely cut off from GPS signals. These systems utilize simultaneous localization and mapping [SLAM] algorithms to actively identify obstacles, dodge power lines, and execute structural data collection completely independently.
- **Beyond Visual Line of Sight [BVLOS] Logistics:** The holy grail of Drone delivery is now a regulatory and operational reality. Automated cargo corridors are scaling across healthcare networks and e-commerce supply chains. BVLOS operations allow heavy-lift Drones to Fly predetermined regional routes spanning dozens of kilometers autonomously, monitored Remotely by a central captain sitting in a distant operations center.
- **Drone Swarm Technology:** Moving from a single Aircraft to coordinated groups, swarm intelligence allows multiple Drones to communicate with each other in real time, distributing tasks efficiently across a single mission. A single operator can deploy a swarm of twenty Drones to execute a massive, multi-acre search and rescue operation or a complex agricultural survey in a fraction of the time required by a standalone Aircraft.
- **Hydrogen Fuel Cells and Solar Integration:** While lithium-ion batteries remain the dominant commercial standard, hydrogen fuel cell Technology is seeing massive adoption for long-endurance applications. Hydrogen-powered Drones are achieving Flight times exceeding three to four Hours, opening up vast new capabilities for

continuous linear pipeline inspections, maritime border tracking, and wide-area search operations.

## Future Growth Projections

The forward horizon for the unmanned Aviation industry points toward exponential expansion. The global Drone market is definitively projected to breach the 210 billion dollar milestone by the mid-2030s, sustained by a continuous 9.6% annual compounding rate.

Within this global landscape, India is positioned as the fastest-growing regional market in the entire Asia-Pacific territory. Propelled by the government's explicit strategic vision to establish India as a premier global Drone hub by 2030, local manufacturing output, service revenues, and certified Pilot Pilot hiring are expected to maintain an aggressive compounding growth rate of over 14.6%.

As the hardware costs of Drones continue to decline due to localized mass production under the PLI scheme, the financial value of the industry will increasingly concentrate in enterprise software analytics, autonomous traffic management infrastructure [UTM], and specialized mission execution. For the certified Remote Pilot in Command, this growth translates into a massive, sustained demand for professional Airmanship. The industry is rapidly moving toward a future where automated Drones are seamlessly integrated into everyday civil Airspace, turning the role of the professional Drone Captain into one of the most vital, high-earning Technical professions of the modern digital economy.

# Chapter 2A Drone Industry Overview

## Global Market Size and Economic Impact

The unmanned Aviation sector has officially transitioned from an emerging niche into a driving pillar of the global aerospace economy. The total global Drone market—spanning military, commercial, and consumer hardware, software, and services—is valued at approximately 100.74 billion dollars. Driven by rapid automation, regulatory ease, and deep integration into industrial workflows, the market is on a trajectory to expand to over 210 billion dollars by the early 2030s, sustaining a steady compound annual growth rate [CAGR] of nearly 9.6%.

Within this overarching footprint, the commercial Drone segment is experiencing the fastest rate of capital injection. Valued at 28 billion dollars, the commercial sector is expanding at an explosive annual rate of more than 9.3%. Regionally, the Asia-Pacific territory has emerged as the dominant powerhouse, capturing over 31.6% of global revenue. This leadership is fueled by intense industrial digitization, massive infrastructure projects, and large-scale agricultural adoption across developing economies.

From an operational standpoint, hardware continues to command the largest share of market value at roughly 61%, led by the mass manufacturing of rotary-blade systems which hold 78% of the market due to their unique hovering capabilities. However, the true high-margin frontier is shifting rapidly toward Drone-as-a-Service [DaaS] business models, cloud-based telemetry analytics, and hybrid VTOL platforms, which grew by more than 22% over the past year alone.

## The Indian Drone Ecosystem

India represents one of the most dynamic and aggressively growing Drone ecosystems in the world. Once constrained by tightly restrictive import laws and manual multi-agency approval structures, the domestic market has completely opened up. Today, India's commercial Drone sector alone is valued at 1.2 billion dollars, with projections positioning it to double to 2.4 billion dollars over the next several years, reflecting a robust growth trajectory. When factoring in the heavy modernization of defense and border reconnaissance platforms, the total Indian UAV market is expected to cross 1.8 billion dollars.

The architecture of India's ecosystem is unique because it is heavily driven by a standardized, digital-first infrastructure. Rather than scaling via fragmented, unregulated private networks, the Indian Drone industry is anchored firmly around the eGCA and Digital Sky frameworks. This structural centralization has turned the country into a fertile laboratory for automated Technology. The domestic ecosystem now features over 38,000 registered Drones holding valid Unique Identification Numbers, an active operational workforce of nearly 40,000 certified Remote Pilots, and a rapidly expanding training foundation consisting of more than 240 DGCA-approved Remote Pilot Training Organisations nationwide.

## Transformative Government Initiatives

The rapid scaling of India's Drone sector is the direct result of deliberate, coordinated legislative policies designed to eliminate operational friction and foster self-reliance. The transition began with the liberalization of the Drone Rules, which replaced over 90% of traditional compliance forms with a trust-based electronic filing system and reduced uniform processing fees to nominal levels. This baseline was further strengthened by a flat rationalization of the Goods and Services Tax [GST] to 5% for all certified Drone systems.

To catalyze domestic manufacturing and eliminate reliance on foreign supply chains, the Ministry of Civil Aviation instituted the Production Linked Incentive [PLI] scheme for Drones and Drone components. This policy provides direct financial incentives to local manufacturers, sparking massive domestic research into localized Flight controllers, propulsion motors, and carbon-fiber composite manufacturing. This is closely pAired with a strict import ban on foreign-assembled conventional

Drones, ensuring that public and private capital flows directly into Indian manufacturing pipelines.

On the application front, two massive public programs are driving nationwide adoption:

- **The SVAMITVA Scheme:** A monumental land-mapping project executed by the Ministry of Panchayati Raj. Utilizing high-precision mapping Drones equipped with RTK navigation, the initiative has successfully surveyed over three lakh villages across India, creating highly accurate digital property cards and providing secure asset ownership documentation to millions of rural citizens.
- **The Namo Drone Didi Scheme:** A major social and agricultural empowerment initiative that has deployed thousands of advanced agricultural spraying Drones directly to women-led rural self-help groups. This project has formalized Drone-based precision pesticide and fertilizer application at scale, turning rural women into high-earning, certified Drone captains and transforming traditional farming methods.

## The Startup Landscape and Capital Inflow

India's Drone startup landscape has moved decisively past the proof-of-concept phase, evolving into a highly competitive network of publicly listed enterprises and heavily funded deep-Tech engineering firms. Startups are no longer merely assembling imported parts; they are securing intellectual property rights for custom Flight controller algorithms, robust anti-Drone defense mechanisms, and highly efficient structural Airframes.

Pioneering firms like IdeaForge Technology have established immense market dominance, scaling from early-stage academic spin-offs into major defense and enterprise suppliers capable of listing publicly on national stock exchanges. Alongside them, companies like Asteria Aerospace, Aarav Unmanned Systems [AUS], and Garuda Aerospace have successfully commercialized enterprise-grade mapping and agricultural software solutions.

Concurrently, deep-Tech startups like Detect Technologies have integrated industrial Drone workflows with automated machine learning pipelines, allowing automated Drones to spot microscopic structural defects, pipeline leaks, and corrosion across massive oil refineries and power plants without human intervention. This robust startup environment has attracted significant venture capital, with early-stage funding increasingly focusing on specialized payload development, long-endurance propulsion systems, and Drone logistics infrastructure.

## Emerging Technological Trends

The state of uncrewed Aviation is being continuously reshaped by the convergence of automation, edge computing, and alternative energy sources. As a professional Drone Captain, keeping pace with these emerging trends is essential for maintaining a competitive edge in the commercial job market.

- **Artificial Intelligence and Autonomous Navigation:** Modern Drones are moving away from continuous manual radio guidance. Integrated with Edge AI processors and advanced computer vision, 2026-generation Drones can navigate safely through dense, complex environments even when completely cut off from GPS signals. These systems utilize simultaneous localization and mapping [SLAM] algorithms to actively identify obstacles, dodge power lines, and execute structural data collection completely independently.
- **Beyond Visual Line of Sight [BVLOS] Logistics:** The holy grail of Drone delivery is now a regulatory and operational reality. Automated cargo corridors are scaling across healthcare networks and e-commerce supply chains. BVLOS operations allow heavy-lift Drones to Fly predetermined regional routes spanning dozens of kilometers autonomously, monitored Remotely by a central captain sitting in a distant operations center.
- **Drone Swarm Technology:** Moving from a single Aircraft to coordinated groups, swarm intelligence allows multiple Drones

to communicate with each other in real time, distributing tasks efficiently across a single mission. A single operator can deploy a swarm of twenty Drones to execute a massive, multi-acre search and rescue operation or a complex agricultural survey in a fraction of the time required by a standalone Aircraft.

- **Hydrogen Fuel Cells and Solar Integration:** While lithium-ion batteries remain the dominant commercial standard, hydrogen fuel cell Technology is seeing massive adoption for long-endurance applications. Hydrogen-powered Drones are achieving Flight times exceeding three to four Hours, opening up vast new capabilities for continuous linear pipeline inspections, maritime border tracking, and wide-area search operations.

## Future Growth Projections

The forward horizon for the unmanned Aviation industry points toward exponential expansion. The global Drone market is definitively projected to breach the 210 billion dollar milestone by the mid-2030s, sustained by a continuous 9.6% annual compounding rate.

Within this global landscape, India is positioned as the fastest-growing regional market in the entire Asia-Pacific territory. Propelled by the government's explicit strategic vision to establish India as a premier global Drone hub by 2030, local manufacturing output, service revenues, and certified Pilot Pilot hiring are expected to maintain an aggressive compounding growth rate of over 14.6%.

As the hardware costs of Drones continue to decline due to localized mass production under the PLI scheme, the financial value of the industry will increasingly concentrate in enterprise software analytics, autonomous traffic management infrastructure [UTM], and specialized mission execution. For the certified Remote Pilot in Command, this growth translates into a massive, sustained demand for professional Airmanship. The industry is rapidly moving toward a future where automated Drones are seamlessly integrated into everyday civil Airspace, turning the role of the professional Drone Captain into one of the most vital, high-earning Technical professions of the modern digital economy.

- **The Namo Drone Didi Scheme:** A major social and agricultural empowerment initiative that has deployed thousands of advanced agricultural spraying Drones directly to women-led rural self-help groups. This project has formalized Drone-based precision pesticide and fertilizer application at scale, turning rural women into high-earning, certified Drone captains and transforming traditional farming methods.

# Chapter 3: Understanding Drone Terminology

- **Aviation Terms**
- **Drone-Specific Terms**
- **Technical Acronyms**
- **Industry Vocabulary**

# Chapter 3 Understanding Drone Terminology

To communicate effectively inside a commercial control room, during an Airport tower briefing, or within a high-stakes corporate enterprise proposal, you must move past colloquial consumer slang. A professional Drone Captain must speak a unified language that bridges the gap between traditional heritage Aviation and digital cutting-edge robotics.

This chapter establishes your foundational industry vocabulary, breaking down essential Aviation metrics, Drone-specific parameters, and Technical acronyms that every operator must know by heart.

## Heritage Aviation Terms

When an unmanned system enters civil Airspace, it becomes subject to the exact same atmospheric laws and communication standards that govern traditional commercial Aircraft. You must master these five core heritage Aviation metrics to maintain professional situational awareness.

### 1. Altitude Metrics: AGL vs. MSL

Understanding how your height is measured can be the difference between a flawless Flight and an immediate Airspace violation.

- **AGL** [Above Ground Level]: This refers strictly to the vertical distance between your Aircraft and the physical ground directly underneath it. If your Drone is hovering 100 feet above a 400-foot hill, your AGL altitude is exactly 100 feet. DGCA and global regulatory operational ceilings are almost always mandated in AGL.
- **MSL** [Mean Sea Level]: This measures the vertical distance of the Aircraft above the standard sea-level geoid, completely independent of the terrain below. If you take off in a mountainous region like Manali, your ground elevation might already be 6,000

feet MSL. Manned Aircraft Pilots look at MSL altimeters; you must understand both to ensure you maintain proper vertical separation from low-Flying manned traffic.

**2. The Three Axes of Flight**

Every Aircraft maneuvers through three-dimensional space by rotating around three mutually perpendicular axes, intersecting at the vehicle's exact center of gravity.

- **Pitch:** The up-and-down movement of the Aircraft's nose, rotating along its lateral axis [running from wingtip to wingtip or side-to-side across the frame].
- **Roll:** The banking or tilting movement of the Aircraft to the left or right, rotating along its longitudinal axis [running straight from the nose to the tail].
- **Yaw:** The side-to-side directional change of the Aircraft's nose, rotating along its vertical axis [running straight up and down through the center of the chassis].

3. Airfoil and Angle of Attack [AoA]

An Airfoil is any cross-sectional surface—such as a fixed wing or a spinning propeller blade—specifically designed to generate aerodynamic lift when interacting with an oncoming stream of Air. The imaginary straight line running directly from the leading edge to the trailing edge of this shape is called the chord line. The acute angle formed between this chord line and the direction of the relative wind is the Angle of Attack. If the AoA becomes too steep, the smooth Airflow over the surface breaks away, causing an immediate loss of lift known as an aerodynamic stall.

4. **METAR and TAF** Weather Reporting

Before executing any commercial Flight, a Pilot must check official Aviation weather bulletins rather than standard consumer smartphone apps.

- **METAR** [Meteorological Aerodrome Report]: A highly structured, coded text report issued at regular Hourly intervals by Airport weather stations, detailing real-time surface wind velocities, cloud ceilings, temperatures, and barometric pressure settings.

- TAF [Terminal Aerodrome Forecast]: A forward-looking meteorological forecast detailing expected atmospheric changes, low-level wind shear, and visibility limitations over a specific twenty-four to thirty-six Hour window within a designated Airport terminal radius.

5. **NOTAM** [Notice to Airmissions]

A NOTAM is an urgent time-sensitive alert issued by civil Aviation authorities to notify Pilots of temporary conditions, hazards, or operational changes within a specific Airspace volume. This includes temporary Flight restrictions around VIP movements, military training exercises, broken ground-based navigation beacons, or sudden localized hazard zones. Checking active NOTAMs is a mandatory element of every pre-Flight briefing.

## Drone-Specific Operational Terms

Unmanned Aircraft systems rely on highly specialized workflows and digital components that have no direct structural equivalent in traditional heritage cockpits.

### 1. Geofencing

An integrated safety software feature that utilizes GPS coordinates to construct an invisible, virtual three-dimensional boundary around a designated operational space. If a Drone attempts to breach this pre-configured digital boundary, the onboard Flight controller will automatically halt forward progression, ignore further stick inputs in that direction, or initiate an automated safe landing.

### 2. Return-to-Home [RTH] / Fail-Safe Mode

An automated emergency protocol hardwired into the Drone's firmware. If the wireless data link between the Ground Control Station and the Aircraft is completely severed, or if the onboard battery levels drop past a critical safety threshold, the Drone overrides manual control, ascends to a pre-configured clear altitude, navigates directly back to its exact launch coordinates using GPS tracking, and executes a controlled autonomous landing.

**3. Payload**

The total combined weight of all removable, non-essential hardware attachments carried by the Drone to execute a specific commercial mission. This encompasses digital cameras, multi-spectral mapping sensors, agricultural chemical tanks, thermal imaging devices, or specialized cargo delivery mechanisms. The payload is completely distinct from the Drone's structural operating weight.

4. **Ground Sampling Distance [GSD]**

A critical metric utilized in Aerial mapping, photogrammetry, and surveying. GSD defines the physical distance between the centers of two consecutive pixels measured on the ground. For example, a GSD of 2 centimeters means that a single pixel in your Aerial map represents exactly 2 square centimeters of real-world land area. GSD is determined by a combination of sensor resolution, focal length, and Flight altitude AGL.

5. **Ground Control Points [GCPs]**

Precisely marked, high-contrast targets placed flat on the ground across a mapping site before a Flight. The exact geographical coordinates of these targets are measured using high-precision ground-based GNSS survey rovers. During post-processing, these points are used to anchor your Drone imagery to the earth, compressing absolute map errors down to sub-centimeter accuracies.

## Technical Acronyms

Modern unmanned Aviation is heavily data-driven, relying on a complex web of electronics and wireless infrastructure. You must recognize these Technical acronyms instantly during systems checks.

1. ESC [Electronic Speed Controller]

The specialized electronic circuit responsible for regulating the speed and rotational direction of a Drone's electric brushless motors. The ESC acts as the direct muscular link between the Flight controller and the motors, translating low-voltage digital commands into heavy-ampere three-phase electrical currents thousands of times per second to keep the Aircraft perfectly balanced.

**2. IMU [Inertial Measurement Unit]**

The primary internal sensor cluster hidden inside the Flight controller. An IMU contains a combination of miniature electronic tri-axial accelerometers, gyroscopes, and occasionally magnetometers. It continuously tracks the Drone's acceleration forces, angular rotational changes, and physical orientation, feeding this raw positional data to the central processor to maintain rock-solid Flight stability.

3. telemetry

The continuous, bidirectional wireless transmission of real-time operational data between the Aircraft and the ground control station. The telemetry stream allows the Pilot to continuously monitor the Drone's real-time battery cell voltages, GPS satellite connections, Flight speeds, exact altitudes, heading directions, and critical thermal statuses during execution.

4. RTK and PPK Precision Navigation

- **RTK [Real-Time Kinematic]**: A high-precision satellite navigation Technique that uses a continuous live data link from a local ground-based base station to correct satellite positioning errors in real time as the Drone flies, achieving centimeter-level tracking accuracy.
- **PPK [Post-Processing Kinematic]:** An alternative precision method where positional corrections are not transmitted over a live radio link during the Flight. Instead, the raw satellite data is recorded onboard the Aircraft and later merged and corrected with ground station data after the Drone lands.

5. BVLOS [Beyond Visual Line of Sight]

An advanced operational Flight profile where the Drone is flown over long distances extending far past the Pilot's direct, unaided visual field. BVLOS missions rely entirely on onboard instruments, automated Flight paths, and forward-facing cameras. It is the operational foundation for long-range package logistics, cross-country pipeline security, and multi-kilometer railway mapping.

## Core Industry Vocabulary

To operate fluently within enterprise environments, you must integrate these standard industry phrases into your daily corporate vocabulary.

**1. Airworthiness**

The formal legal and structural status of an Aircraft, confirming that it matches its approved type design and is in a safe condition for operations. In the Drone world, Airworthiness is validated through active pre-Flight maintenance inspections, up-to-date firmware syncs, and the possession of an official DGCA Type Certificate.

**2. Visual Line of Sight [VLOS]**

The foundational safety standard for civil Drone operations. VLOS dictates that the Remote Pilot in Command or an assigned Visual Observer must maintain direct, continuous, unaided visual contact with the Aircraft throughout the entire Flight duration. This contact must be sufficient to monitor the Drone's orientation, detect potential Airborne hazards, and avoid collisions with birds, buildings, or low-Flying manned Aircraft without using binoculars or video screens.

**3. Remote Pilot in Command [RPIC]**

The certified Professional Pilot who holds ultimate legal and operational responsibility for the safe, compliant conduct of an entire Unmanned Aircraft System mission. The RPIC has absolute final authority over the deployment, operation, and emergency termination of the Flight, regardless of whether they are manually operating the sticks or monitoring an automated Flight sequence.

4. Orthomosaic Map

A highly detailed, geometrically corrected photographic map of a landscape created by digitally stitching together hundreds of overlapping

Aerial photos captured by a mapping Drone. An orthomosaic map is corrected for topographic relief, camera lens distortion, and tilt, meaning it can be used to measure true real-world horizontal distances, areas, and precise coordinates directly on a screen.

# PART II

# Drone Types & Technology

# CHAPTER 4: ANATOMY OF A DRONE

- **Airframe**
- **Flight Controller**
- **Motors**
- **ESC**
- **Propellers**
- **GPS**
- **Telemetry System**
- **Antennas**
- **Cameras**
- **Payload Systems**

# Chapter 4 Anatomy of a Drone

## 1. The Airframe: Structural Backbone

The Airframe is the physical chassis or skeleton that houses, protects, and supports every internal electronic component and external payload system. In commercial Aviation, Airframes are constructed using advanced aerospace-grade materials like carbon fiber, Kevlar, or high-tensile aluminum alloys. These materials provide an exceptional strength-to-weight ratio, ensuring the frame can withstand intense aerodynamic torsional forces, high wind resistances, and sudden hard landings while remaining as light as possible to maximize battery efficiency.

The physical configuration of the Airframe directly determines its Flight dynamics. For multi-rotors, the frame consists of a central hub that seals the delicate electronics away from dust and moisture, extending outward into rigid arms that hold the propulsion systems. For fixed-wing systems, the Airframe consists of a aerodynamic fuselage, main lifting wings, and rear tail stabilizers designed to slice smoothly through Air currents.

## 2. The Flight Controller: Onboard Brain

The Flight Controller [FC] is the central processing unit and nervous system of the Drone. It is a highly specialized circuit board packed with powerful microprocessors running real-time Flight management software. The Flight controller's primary job is to keep the Aircraft stable and execute Pilot commands smoothly. It achieves this by continuously reading data from an internal sensor suite known as the Inertial Measurement Unit [IMU], which contains micro-electronic gyroscopes and accelerometers tracking the Drone's position, tilt, and acceleration thousands of times per second.

The Flight controller takes this raw positioning data, calculates the exact adjustments needed to maintain stable Flight, and sends instant corrective signals to the propulsion system. It also acts as the central clearinghouse for all auxiliary data, managing everything from safety geofencing boundaries to autonomous return-to-home protocols.

## 3. Motors: Power and Propulsion

Commercial Drones rely almost exclusively on Brushless DC [BLDC] motors to generate mechanical thrust. Unlike cheap, brushed hobbyist motors that rely on mechanical carbon brushes that wear down from friction, brushless motors utilize an electronic commutation system. A fixed inner ring of electromotive copper coils interacts magnetically with an outer spinning bell lined with powerful rare-earth permanent magnets.

This design completely eliminates friction, allows for massive heat dissipation, reduces electrical noise, and ensures exceptional mechanical reliability over hundreds of Flight Hours. For a professional operator, the key metric to understand is the motor's Kv rating, which dictates the theoretical revolutions per minute [RPM] the motor will spin per one volt of electricity applied under no load. Lower Kv motors spin larger propellers with immense rotational torque for heavy industrial payloads, while high Kv motors spin smaller blades at blinding speeds for agile maneuvering.

## 4. ESC (Electronic Speed Controller): Muscular Translation

The Electronic Speed Controller [ESC] acts as the high-current muscular bridge connecting the main power distribution board, the Flight controller, and the brushless motors. A Drone cannot vary its Flight direction by turning a traditional steering wheel; a multi-rotor changes its orientation by precisely varying the relative rotational speeds of its different motors.

The Flight controller cannot send raw power to a motor directly. Instead, it sends low-voltage digital command signals to the ESC. The ESC takes this data and uses high-power field-effect transistors to rapidly pulse heavy electrical currents from the main battery pack into the motor's three internal phases. By modulating these electrical pulses thousands of times per second, the ESC can instantly change a motor's speed or direction, allowing the Drone to react dynamically to sudden wind shears or aggressive Pilot stick inputs.

## 5. Propellers: Aerodynamic Lift

Propellers are the physical Airfoils responsible for translating the rotational torque of the brushless motors into aerodynamic lift and forward thrust. Propeller blades are twisted at a specific angle or pitch, designed to bite into the oncoming Air mass as they rotate. Propellers are pAired systematically across a Drone frame in two distinct configurations: Clockwise [CW] and Counter-Clockwise [CCW].

This alternating rotational pattern is critical; it cancels out the rotational physics of torque. If all four propellers on a quadcopter spun in the exact same direction, the entire physical body of the Drone would violently spin in the opposite direction. By balancing equal numbers of clockwise and counter-clockwise spinning Airfoils, the Drone remains perfectly stable in the yaw axis. Commercial blades are manufactured from rigid carbon fiber to prevent the blade from flexing or warping under heavy aerodynamic loads.

## 6. GPS / GNSS System: Spatial Positioning

To execute autonomous mapping missions, hover perfectly stable in high winds without drifting, and track return-to-home coordinates accurately, a Drone must know its exact physical coordinates on Earth. This is handled by the onboard Global Navigation Satellite System [GNSS] module, commonly referred to as the GPS. Modern enterprise Drones utilize advanced multi-constellation receivers that track multiple global satellite networks simultaneously, including GPS [United States], GLONASS [Russia], Galileo [Europe], and BeiDou [China].

For industrial-grade mapping, standard GPS coordinates are not accurate enough, as atmospheric distortions can cause positioning errors of up to several meters. Therefore, enterprise systems integrate Real-Time Kinematic [RTK] GNSS receivers. By continuously receiving live, sub-second differential corrections from a fixed ground-based base station or local cellular network, an RTK-enabled Drone can pinpoint its exact spatial location in mid-Air down to sub-centimeter accuracies.

## 7. Telemetry System: Bidirectional Communication

The telemetry system is the wireless communications infrastructure that establishes a continuous, invisible data umbilical cord between the Aircraft in Flight and the Ground Control Station [GCS] on the ground. Telemetry relies on high-bandwidth radio frequency [RF] transceivers operating on dedicated civil Aviation and industrial frequencies, typically around 2.4 GHz, 5.8 GHz, or specialized lower long-range frequencies.

This bidirectional link operates simultaneously on two distinct paths. The downlink transmits real-time Aircraft health diagnostics down to your Pilot screen, displaying critical Flight data such as individual battery cell voltages, live GPS satellite counts, speed over ground, current altitude, heading directions, and any active system error flags. The uplink path simultaneously transmits your manual joystick commands or autonomous waypoint Flight paths from the ground station up to the Drone's Flight controller.

## 8. Antennas: RF Signal Distribution

Antennas are the physical hardware components responsible for converting electrical currents from the onboard radio transmitters into electromagnetic waves traveling through the Air, and vice versa. Without correctly configured and maintained antennas, even the most powerful telemetry system will suffer from immediate signal degradation and catastrophic control link failures.

Drones utilize two primary categories of antennas:

- **Omnidirectional Antennas:** These radiate and receive radio frequency signals in a uniform, 360-degree doughnut-like shape around the hardware. They are installed directly onto the Drone chassis to ensure that no matter how the Aircraft pitches, rolls, or changes its heading relative to the Pilot, a stable wireless connection is maintained.
- **Directional Antennas:** These concentrate their radio energy into a tight, highly focused beam pointing in a specific direction. They are installed on ground-based tracking stations to punch through long distances for long-range beyond visual line of sight missions.

## 9. Cameras: Visual Data Acquisition

While early Drones carried simple consumer action cameras, modern commercial systems deploy advanced, stabilized imaging payloads tailored to specific data verticals.

- **RGB Cameras:** High-resolution visual sensors equipped with large global shutters, used to capture crisp, distortion-free Aerial photographs for real estate modeling, media production, and industrial photogrammetry mapping.
- **Thermal Infrared Cameras:** Radiometric sensors that detect long-wave infrared radiation, mapping relative surface temperatures rather than visible light. These are critical tools for locating microscopic solar panel defects, identifying heat loss in high-voltage power grids, and tracking missing persons during nighttime search operations.
- **Multispectral Sensors:** Specialized cameras that capture specific, narrow wavebands of light, including near-infrared and red-edge frequencies. These sensors allow agricultural captains to look past visual plant color and directly analyze internal chlorophyll levels, mapping crop health variations across thousands of acres.

## 10. Payload Systems and Gimbal Stabilizers

The payload is the reason the Drone leaves the ground; it is the commercial tool that generates value. To ensure that the data captured by a camera or a heavy LiDAR laser scanner is accurate and blur-free, the payload must be isolated from the intense vibrations generated by the high-speed spinning motors and the sudden tilting movements of the Airframe. This stabilization is handled by a motorized Gimbal System.

An enterprise gimbal utilizes an independent internal IMU cluster and ultra-responsive brushless direct-drive motors operating across three distinct axes: pitch, roll, and yaw. If a sudden gust of wind tilts the Drone 15 degrees to the left, the gimbal instantly tilts 15 degrees to the right, keeping the camera sensor perfectly level and locked onto its target. Advanced payloads also feature quick-release modular mounting plates, allowing a Pilot to swap a visual mapping camera for a thermal sensor or a heavy cargo dropping mechanism on the field within seconds.

# CHAPTER 5: TYPES OF DRONES BY DESIGN

## Multi-Rotor Drones

- Tricopter
- Quadcopter
- Hexacopter
- Octocopter

Drone.AsiaticInCorp.org

# Types of Drones by Design

## Multi-Rotor Drones

### Tricopter

- ***Configuration:*** 3 Rotors
- ***Control:*** Requires a tilt mechanism for yaw
- ***Efficiency:*** Less efficient than Quad
- ***Application:*** hobbyist use

### Quadcopter

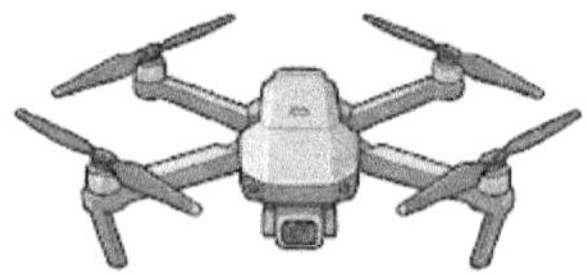

- ***Configuration:*** 4 Rotors
- ***Control:*** Simpler, highly maneuverable
- ***Popularity:*** The most common type
- ***Application:*** Photography, survey, hobby

### Hexacopter

- ***Configuration:*** 6 Rotors
- ***Redundancy:*** Can fly with one failed motor
- ***Payload:*** Higher lift capacity
- ***Application:*** Professional film, heavy sensors

### Octocopter

- ***Configuration:*** 8 Rotors
- ***Max Redundancy:*** Very safe operation
- ***Stability:*** Highest stability and payload
- ***Application:*** Large cameras, complex missions

## Non-Multi-Rotor Types (for perspective)

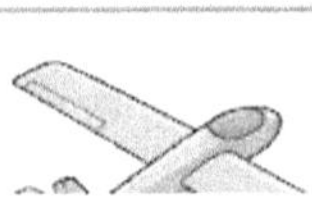

### Fixed-Wing

### Hybrid (VTOL)

# Chapter 5 Types of Drones by Design

The sky is no longer the exclusive domain of birds and manned Aircraft. Over the last few decades, Unmanned Aerial Vehicles [UAVs]—commonly known as Drones—have transitioned from specialized military hardware to ubiquitous tools shaping industries, science, and recreation.

While Drones can be categorized by their range, payload capacity, or autonomy level, the most fundamental way to classify them is by their **structural design**. The design of a Drone dictates how it generates lift, how it maneuvers, how long it can stay Airborne, and ultimately, what kinds of missions it can execute.

In this chapter, we will explore the most popular and versatile category of UAVs: **Multi-Rotor Drones**, focusing specifically on the structural and operational differences between **Tricopters, Quadcopters, Hexacopters, and Octocopters**.

## The Mechanics of Multi-Rotor Drones

Before diving into specific configurations, it is essential to understand how multi-rotor Drones operate. Unlike traditional fixed-wing Aircraft that rely on forward momentum to generate lift across a wing surface, multi-rotor Drones use two or more upward-facing propellers.

## How Multi-Rotors Fly

Multi-rotors vary the relative speed of specific motors to change their orientation and position in three-dimensional space:

- **Pitch [Tipping forward or backward]:** Achieved by speeding up the rear rotors and slowing down the front rotors [or vice versa].

- **Roll [Tipping left or right]:** Achieved by speeding up the rotors on one side and slowing down the ones on the opposite side.
- **Yaw [Rotating clockwise or counterclockwise]:** Achieved by changing the balance of torque. In a standard multi-rotor, half the propellers spin clockwise [CW] and the other half spin counterclockwise [CCW] to cancel out each other's rotational force. Speeding up only the CW rotors causes the Drone's body to yaw counterclockwise.

## General Advantages and Trade-offs

Because they can hover perfectly in place and take off or land vertically [VTOL], multi-rotors do not require a runway. This makes them incredibly agile and perfect for tight, cluttered spaces. While they must constantly expend battery power to stay aloft—which generally limits their Flight times to between 20 and 45 minutes—their mechanical simplicity and hovering capabilities make them the go-to choice for many applications.

## 1. Tricopters: The Agile Three-Rotor Anomalies

As the name implies, a **Tricopter** features three motors, usually arranged in a "Y" or "T" shape, with two rotors at the front pointing outward and one rotor at the rear.

### The Mechanics of the Tricopter

Tricopters present a unique mechanical engineering challenge. With an odd number of rotors, a tricopter cannot rely purely on varying motor speeds to control its yaw [rotation]. To solve this, a tricopter requires a **mechanical servo motor** attached to the rear rotor. This servo physically tilts the rear motor left or right, vectoring the thrust to control the Drone's yaw.

### Key Advantages of Tricopters

- **Saves on Costs:** Fewer motors and electronic speed controllers [ESCs] mean lower initial component costs.

- **Dynamic, Swooping Flight:** Because of the physical tilting mechanism, tricopters bank into turns much like a fighter jet or a bird, offering a highly unique, organic Flying experience.
- **Foldable/Compact Designs:** The "Y" shape allows the front arms to fold backward easily for transport.

**Best Used For:** Hobbyists, DIY Drone builders, and RC enthusiasts looking for an engaging, unique Flight dynamic rather than a commercial workhorse.

## 2. Quadcopters: The Universal Standard

The **Quadcopter** is the undisputed king of the Drone world. From cheap toy Drones to multi-thousand-dollar filmmaking rigs, the four-rotor design is the blueprint for modern UAVs.

**The Mechanics of the Quadcopter**

Quadcopters utilize a beautifully simple layout, typically configured in an "X" or "+" shape. Two diagonally opposite motors spin clockwise [CW], while the other two spin counterclockwise [CCW]. This layout perfectly balances the aerodynamic torque, eliminating the need for any complex mechanical servos like those found in tricopters. All maneuvering is done purely by digitally adjusting the RPM of the individual brushless motors.

**Key Advantages of Quadcopters**

- **Simplicity and Reliability:** With no moving parts other than the motors themselves, quadcopters are incredibly durable and easy to maintain.
- **Massive Ecosystem:** Because of their popularity, parts, Flight software, and accessories are highly accessible and affordable.
- **Optimized Efficiency:** Striking the perfect balance between lift capacity and battery consumption, quadcopters usually offer excellent Flight times among multi-rotors.

**Best Used For:** Aerial photography, consumer videography, First-Person View [FPV] Drone racing, mapping, and general consumer recreation.

## 3. Hexacopters: Stepping Into Professional Utility

When a Drone needs to lift something heavier or Fly over populated areas where safety is paramount, four rotors are often not enough. Enter the **Hexacopter**, which utilizes six motors arranged symmetrically in a circular wheel, typically spaced 60 degrees apart.

### The Mechanics of the Hexacopter

Like the quadcopter, the hexacopter balances torque by splitting its propulsion evenly: three motors spin CW, and three spin CCW. The addition of two extra motors drastically changes the Drone's capability profile, particularly regarding power and redundancy.

### Key Advantages of Hexacopters

- **Built-in Redundancy:** If a single motor fails on a hexacopter, the Flight controller can redistribute power to the remaining five motors, allowing the Pilot to make a controlled emergency landing.
- **Higher Payload Capacity:** Six motors generate significantly more thrust, allowing the Drone to carry heavier professional cameras, LiDAR sensors, or thermal equipment.
- **Smoother Flight:** The distributed thrust creates a more stable platform that behaves exceptionally well in high-wind conditions.

**Best Used For:** Professional cinematography, structural inspections, surveying, agricultural crop scouting, and carrying mid-tier sensory payloads.

## 4. Octocopters: Ultimate Power and Redundancy

At the absolute apex of the multi-rotor hierarchy sits the **Octocopter**. Featuring eight independent rotors, these massive machines are designed strictly for heavy industrial and high-stakes commercial operations.

### The Mechanics of the Octocopter

Octocopters generally come in two configurations:

1. **Flat 8:** Eight arms radiating outward from a central hub, each carrying a single motor.

2. **Coaxial X8:** An "X" shaped frame with four arms, but each arm features two motors mounted back-to-back [one pushing upward, one pushing downward]. This configuration keeps the Drone's footprint small while maintaining eight points of thrust.

With four motors spinning CW and four spinning CCW, the sheer volume of Air displaced by an octocopter gives it unmatched stability.

**Key Advantages of Octocopters**

- **Extreme Redundancy:** An octocopter can lose two, or sometimes even three motors [depending on the layout and payload weight] and still maintain stable Flight.
- **Maximum Payload Capacity:** These Drones are engineered to lift heavy, specialized equipment, including Hollywood-grade IMAX cameras, industrial LiDAR scanners, or cargo packages.
- **Unrivaled Stability:** The high motor density allows the Flight controller to make micro-adjustments at lightning speed, holding its position perfectly even in severe gale-force winds.

**Best Used For:**

High-end Hollywood filmmaking, heavy-lift logistics and cargo delivery, search and rescue operations carrying specialized life-saving gear, and advanced industrial mapping.

This section explores the mechanics and practical use-cases of Fixed-Wing platforms, Single and Coaxial rotor designs, and the modern emergence of Hybrid VTOL systems.

## 1. Fixed-Wing Drones: The Long-Distance Cruisers

Fixed-wing Drones ditch the multi-motor lifting array entirely, utilizing a rigid wing structure identical to traditional Airplanes.

**Mechanics of Flight**

Fixed-wing Aircraft do not use energy to fight gravity directly. Instead, a rear-mounted or front-mounted motor generates forward propulsion. As Air moves quickly over the curved wing profile, it creates an aerodynamic pressure differential—known as lift—which holds the Aircraft aloft. Control surfaces on the wings and tail [ailerons, elevators, and rudders] tilt slightly to alter Airflow and navigate the sky.

**Survey Drones**

For mapping large areas, fixed-wing variants are highly effective survey tools. They carry advanced photogrammetry sensors, multispectral cameras, or high-definition mapping equipment over massive properties in a single Flight. They can traverse vast geographic expanses line-by-line, capturing perfectly uniform imagery from a fixed altitude.

**Long-Endurance Drones**

Because they rely on natural gliding physics rather than constant, brute-force motorized lifting, fixed-wing designs are excellent for long endurance. With battery configurations or small internal combustion engines, these systems easily sustain Flight times ranging from 1 to several Hours. They can cover hundreds of linear kilometers per deployment, tracking pipelines, examining national borders, or observing wildlife preserves across large swathes of land.

**Key Advantages**

- **Exceptional Range and Efficiency:** Gliding physics allow these platforms to cover significantly more ground per watt of power than any multi-rotor.
- **Stable High-Altitude Flight:** Fixed-wing physics naturally handle high wind currents at altitude, making them reliable platforms for high-resolution landscape analysis.

**Best Used For:** Large-scale agricultural mapping, pipeline/powerline corridor inspection, environmental conservation research, and Aerial surveying.

## 2. Single Rotor Drones: Traditional Helicopter Architecture

A single-rotor Drone is essentially a scaled-down, uncrewed helicopter. It features one massive main rotor on top and a smaller vertical rotor on the tail.

## The Mechanics of Single Rotors

Single rotor systems generate lift using an intricate mechanical assembly called a **swashplate**. Instead of varying motor speeds to move, the main motor spins at a relatively constant rate, and the swashplate alters the physical pitch [blade angle] of the spinning rotors as they rotate.

Because the spinning main rotor generates massive rotational force [torque] that would naturally cause the Drone's body to spin wildly in the opposite direction, the small tail rotor pushes Air sideways to neutralize that twisting force and control the Drone's yaw.

**Key Advantages**

- **High Mechanical Efficiency:** One large rotor blade moves Air far more efficiently than four to eight tiny propellers spinning at ultra-high speeds, allowing for a stronger combination of Flight times and payload capacities.
- **Gas/Hybrid Powertrain Compatibility:** They easily pAir with gas-powered combustion engines, allowing for extended missions without relying entirely on modern battery limits.

**Best Used For:** Heavy-lift agricultural spraying, Remote cargo transit, and operations where hovering is required but payload requirements exceed normal quadcopter limits.

## 3. Coaxial Rotor Drones: Stacked Precision

Coaxial Drones solve the torque challenge of a single rotor by stacking two independent main rotors directly on top of each other along a single vertical drive shaft.

**The Mechanics of Coaxial Flight**

In a coaxial configuration, the two stacked rotors spin in completely opposite directions. The top blade spins clockwise while the bottom blade spins counterclockwise. Because the torque generated by each rotor cancels out the other, the Drone remains stable without needing a tail rotor assembly. Yaw is controlled by subtly shifting the relative speed or pitch of one blade stack over the other.

**Key Advantages**

- **Ultra-Compact Footprint:** Removing the tail boom entirely allows the system to remain small, making it easier to transport and safer to operate near walls, trees, or tight industrial scaffolding.
- **Massive Lifting Force:** Stacking the lifting power directly underneath the central column gives these systems impressive vertical stability and weight tolerance relative to their physical dimensions.

**Best Used For:** Close-range military reconnaissance, tactical situational awareness, and operations in dense, confined spaces where space is limited.

## 4. Hybrid VTOL Drones: The Best of Both Worlds

Hybrid VTOL [Vertical Takeoff and Landing] Drones represent a modern evolutionary step in UAV design, merging the vertical agility of a multi-rotor with the high-speed cruising efficiency of a fixed-wing Airplane.

**The Mechanics of Hybrid Flight**

Hybrid VTOLs feature a split personality. They carry a set of vertical lift rotors [like a quadcopter] alongside a traditional fixed-wing layout with a dedicated forward pusher prop.

1. **Takeoff:** The Drone uses its vertical rotors to lift cleanly off the ground like a helicopter—no runway or launch catapult required.
2. **Transition:** Once at a safe altitude, the forward cruise motor kicks in, pulling the Aircraft forward until Air moving across the wings generates enough independent lift.
3. **Cruise:** The vertical rotors shut off completely, and the Drone cruises efficiently like an Airplane.
4. **Landing:** Upon reaching its destination, the process reverses to allow for a soft, precise vertical touchdown.

**Key Advantages**

- **No Launch Gear Required:** They completely eliminate the need for dangerous catapult launchers or unassisted belly-landings on rough terrain.

- **Industrial Utility Scale:** By combining a simple deployment process with long-range Flight profiles, these hybrid units can execute high-tier mapping, defense, and long-range package transport missions with minimal logistics.

**Best Used For:**

Long-range offshore delivery, multi-thousand-acre mapping projects, border patrol operations, and medical supply transport to Remote regions.

# CHAPTER 6: TYPES OF DRONES BY WEIGHT CATEGORY

- Nano Drones
- Micro Drones
- Small Drones
- Medium Drones
- Large Drones

# Chapter 6 Types of Drones by Weight Category

While a Drone's structural configuration—such as fixed-wing or multi-rotor—dictates how it flies, its total weight determines how it is regulated, where it can legally operate, and the level of risk it poses to shared Airspace. Globally, civil Aviation authorities utilize Maximum Takeoff Mass [MTOM] or All-Up Weight [AUW] as the primary metrics to categorize unmanned Aircraft systems [UAS].

Categorizing Drones by weight provides a clear framework for everything from casual consumer safety to high-stakes industrial Flights. This chapter breaks down the five primary weight classifications: **Nano, Micro, Small, Medium, and Large Drones**, examining their Technical capabilities, practical applications, and general regulatory profiles.

## 1. Nano Drones (Under 250 Grams)

Nano Drones, often referred to as ultralight or toy-grade Drones, represent the smallest and safest tier of unmanned Aircraft. The 250-gram threshold is a universally recognized boundary in global Aviation law.

**Technical Characteristics**

These miniature devices easily fit into the palm of a hand. Because they weigh less than a standard smartphone, they possess very little kinetic energy during Flight. This minimal impact hazard means they rarely require complex structural testing or heavy safety redundancies. While early generations were limited to indoor toy use, modern nano Drones feature highly advanced, stabilized 4K cameras, GPS mapping chips, and intelligent tracking software.

**Primary Applications**

- **Casual Recreation and Vlogging:** Perfect for travelers and content creators due to their extreme portability.

- **Indoor Inspections:** Highly useful for Flying inside tight, enclosed spaces like warehouses, storage tanks, or historical buildings where larger Aircraft pose a structural risk.
- **Hobbyist FPV Racing:** Ultra-lightweight "Tiny Whoop" Drones allow Pilots to safely practice high-speed maneuvers indoors.

**Regulatory Profile**

Because they present minimal risk to public safety, nano Drones enjoy the most relaxed operational boundaries. In many major regulatory frameworks, recreational nano platforms are exempt from mandatory registration and hardware-broadcasting laws like Remote ID. However, if used for commercial, revenue-generating operations, Pilots must still obtain standard commercial certifications.

## 2. Micro Drones (250 Grams to 2 Kilograms)

The micro Drone tier bridges the gap between casual consumer hobby Tech and entry-level professional equipment.

**Technical Characteristics**

Drones in this category weigh between 250g and 2kg. They feature robust brushless motors, large battery bays, and advanced wind-resistance capabilities, making them highly reliable for outdoor operation. This class includes standard consumer platforms used globally for Aerial filmmaking, as well as custom-built 5-inch First-Person View [FPV] freestyle quads.

**Primary Applications**

- **Professional Videography:** The standard choice for real estate marketing, wedding filmmaking, and local news gathering.
- **Close-Range Mapping:** Ideal for small-scale land surveying, construction site updates, and 3D orthomosaic modeling.
- **First-Responder Situational Awareness:** Easily deployed from patrol vehicles to provide rapid overhead views during active traffic incidents or search operations.

### Regulatory Profile

Crossing the 250-gram threshold triggers a significant shift in legal oversight. In nearly all jurisdictions, micro Drones must be registered with national Aviation databases, and they are required to broadcast their identification data wirelessly via Remote ID modules. While recreational users can Fly them with basic online safety certificates, commercial operations require a formal Remote Pilot license.

## 3. Small Drones (2 Kilograms to 25 Kilograms)

Small Drones transition completely into professional, industrial, and enterprise operations. These systems are heavy, powerful, and built to handle specialized corporate workloads.

### Technical Characteristics

Weighing between 2kg and 25kg, small Drones feature expansive multi-rotor frames or medium-sized fixed-wing architectures. They are designed to carry interchangeable, modular payloads rather than fixed consumer cameras. These payloads include high-end cinematic cameras, thermal imaging sensors, and optical gas-imaging equipment. They also possess advanced collision-avoidance arrays and dual-battery systems for enhanced reliability.

### Primary Applications

- **Industrial Infrastructure Inspection:** Used to inspect high-voltage power lines, wind turbines, cell towers, and bridge understructures.
- **Precision Agriculture:** Equipped with multispectral sensors to scan thousands of acres of crops, mapping soil moisture levels and plant health indicators.
- **High-End Cinematography:** Utilized on professional film sets to carry mid-tier cinema cameras and heavy anamorphic lenses.

### Regulatory Profile

Small Drones are strictly regulated because an out-of-control 20-kilogram Aircraft can cause severe property damage or personal injury. Pilots must hold active commercial credentials to operate them. Flying a small Drone over unprotected crowds or outside the Pilot's direct line of sight typically requires rigorous risk assessments and explicit waivers from civil Aviation authorities.

## 4. Medium Drones (25 Kilograms to 150 Kilograms)

Medium Drones leave the consumer market behind entirely. These high-capacity, industrial platforms are engineered for heavy lifting and long-range utility missions.

**Technical Characteristics**

Weighing between 25kg and 150kg, these massive units are often powered by heavy-duty hybrid gas-electric powertrains or high-voltage industrial battery systems. Due to their immense weight, they generate tremendous thrust and feature rugged, heavy-duty landing gear. Many medium Drones utilize complex autonomous Flight paths managed by industrial ground control stations rather than traditional handheld Remote controls.

**Primary Applications**

- **Agricultural Crop Spraying:** The dominant workhorse for modern precision farming. They carry large liquid payloads to automate the application of fertilizers, pesticides, and seeds over vast fields.
- **Cargo Logistics and Delivery:** Used by logistics companies to transport heavy medical supplies, critical tools, or commercial packages across Remote, rugged terrain.
- **Large-Scale Surveying:** Heavy fixed-wing or hybrid VTOL variants carry advanced LiDAR sensors to chart dense forests, coastlines, and expansive mining operations.

**Regulatory Profile**

Operating a medium Drone is legally similar to Piloting a light crewed Aircraft. Drones in this tier require formal Airworthiness certifications and type ratings. Operators must secure specialized commercial permits, establish strict operational safety boundaries, and clear Flights through regional Air traffic control networks.

## 5. Large Drones [Above 150 Kilograms]

Large Drones represent the absolute apex of unmanned Aviation Tech. These massive autonomous Aircraft operate at altitudes and distances comparable to traditional crewed Airplanes and helicopters.

### Technical Characteristics

Weighing over 150kg, large Drones are built using Aviation-grade carbon composites, aluminum alloys, and high-output internal combustion or jet propulsion systems. They feature sophisticated military-grade communication arrays, satellite navigation relays, and full redundancy across all internal Flight computers.

### Primary Applications

- **Strategic Military Operations:** Used for long-endurance border reconnaissance, tactical surveillance, and high-altitude defense missions.
- **Next-Generation Cargo Transit:** Massive autonomous freight platforms designed to transport hundreds of pounds of cargo between dedicated distribution hubs.
- **Urban Air Mobility [UAM]:** The foundational hardware for emerging autonomous Air taxis and passenger-carrying **eVTOL** [electric Vertical Takeoff and Landing] transit systems.

### Regulatory Profile

Large Drones operate inside the highest tiers of regulated civil and military Airspace. They are subject to rigorous Airworthiness certification pathways, mandatory maintenance schedules, and comprehensive operator licensing protocols that mirror commercial Airline standards. Every deployment requires direct Flight plan integration with national Air traffic management systems.

## Summary of Weight Classifications

The Drone weight hierarchy establishes a clear progression from low-risk recreational gadgets to high-capacity industrial machines:

- **Nano [<250g]:** Ultralight, portable, and subject to minimal regulatory restrictions; ideal for hobbyists and indoor Flight.
- **Micro [250g–2kg]:**
- The standard choice for prosumer Aerial photography and casual FPV Flying; requires basic registration and Remote ID.
- **Small [2kg–25kg]:**
- Heavy-duty enterprise tools built for infrastructure inspection, professional filmmaking, and precise environmental mapping.
- **Medium [25kg–150kg]:**
- Industrial machinery optimized for high-volume agricultural spraying and long-range commercial cargo logistics.
- **Large [>150kg]:**
- High-altitude, high-capacity autonomous Aircraft designed for heavy freight, defense operations, and future passenger transit.

# CHAPTER 7: TYPES OF DRONES BY PURPOSE

- **Agricultural Drones**
- **Mapping Drones**
- **Survey Drones**
- **Inspection Drones**
- **Delivery Drones**
- **Defense Drones**
- **Surveillance Drones**
- **Search & Rescue Drones**
- **Cinematography Drones**
- **Photography Drones**
- **Racing Drones**

# Chapter 7 Types of Drones by Purpose

While engineering choices like rotor layouts and weight limits define how an uncrewed system operates, the end application dictates the layout of its physical payload, camera configuration, and internal software stack. Modern Drones are highly specialized tools built for specific industrial, tactical, or creative outcomes.

This chapter categorizes unmanned Aerial vehicles [UAVs] by their intended purpose, analyzing the distinct operational roles they perform across modern commercial, civilian, and defense sectors.

## 1. Commercial and Industrial Workhorses

Industrial platforms are designed for heavy workloads, precise data collection, and physical utility, replacing slower, costlier, and more dangerous manual methods.

**Agricultural Drones**

Agricultural Drones optimize crop management, field scouting, and chemical applications. These systems are split into two core types. Imaging and scouting Drones equipped with multispectral and Normalized Difference Vegetation Index [NDVI] sensors detect plant stress, soil variation, and nutrient deficiencies long before they are visible to the human eye. Spraying and spreading Drones are large multi-rotors capable of carrying heavy liquid payloads. Guided by GPS and RTK [Real-Time Kinematic] positioning, they apply pesticides, fertilizers, and seeds with surgical precision, reducing chemical waste and minimizing ground soil compaction.

**Mapping Drones**

Mapping platforms generate highly accurate, measurable representations of the earth's surface. Using high-resolution RGB cameras and specialized automated Flight paths, they capture a series of overlapping Aerial photographs. These images are processed using photogrammetry software to build detailed 2D orthomosaic maps, digital elevation models [DEMs], and point clouds for town planning, resource management, and forestry.

**Survey Drones**

While mapping Drones focus on visual rendering, survey Drones prioritize extreme spatial accuracy, often down to the millimeter. They carry active payload systems like LiDAR [Light Detection and Ranging] lasers, which pierce dense forest canopies to profile the true topography of the ground underneath. They are vital for civil engineering, open-pit mining volume calculations, and route planning for major highways.

**Inspection Drones**

Built to operate near hazardous, high-value structures, inspection Drones feature advanced omnidirectional obstacle avoidance, shielded electronics, and high-optical-zoom or thermal cameras. They routinely inspect live high-voltage power grids, offshore wind turbines, refinery flare stacks, and bridge understructures, keeping human inspectors safely on the ground.

**Delivery Drones**

Delivery UAVs focus on automated logistics, payload security, and rapid transit. Utilizing fixed-wing or hybrid VTOL designs, these systems travel along pre-programmed corridors to deliver high-priority packages. They feature secure, climate-controlled internal drop-bays, automated package release mechanisms, and emergency parachutes to protect valuable cargo like medical supplies, blood plasma, and organs.

## 2. Tactical, Safety, and Defense Systems

Drones built for defense and public safety prioritize encrypted communication relays, all-weather reliability, long range, and rapid deployment.

**Defense Drones**

Defense platforms span from small, hand-launched reconnaissance units to massive, high-altitude long-endurance [HALE] combat systems. Built with Aviation-grade materials and protected by anti-jamming GPS Tech, they carry specialized radar systems, electronic warfare modules, or laser target designators to execute long-range military strategies without risking crewed Aircraft.

**Surveillance Drones**

Surveillance systems maintain continuous situational awareness over defined borders, critical facilities, or maritime sectors. Engineered for minimal acoustic footprints, they use high-power optical payloads with stabilizing gimbals to track vehicular or human movement from miles away, feeding live encrypted video back to a central command post.

**Search & Rescue Drones**

Search and Rescue [SAR] platforms are rapid-response tools deployed during natural disasters, missing persons cases, or industrial accidents. They feature integrated dual-sensor payloads containing an optical camera alongside a high-sensitivity thermal imaging core. This allows them to quickly spot the heat signature of a lost hiker at night or find survivors trapped beneath debris fields. They can also carry external speakers to broadcast instructions or drop emergency survival kits.

## 3. Creative and Recreational Media Platforms

Media Drones are built to position high-fidelity camera sensors precisely in three-dimensional space, opening up new perspectives for artists, hobbyists, and professional broadcasters.

**Cinematography Drones**

Cinematography platforms are heavy-lift multi-rotors built for professional movie sets and television broadcasts. Rather than using basic integrated camera sensors, they feature heavy-duty three-axis gimbals designed to carry professional cinema cameras and heavy lenses. They typically require a two-person crew consisting of one Pilot to guide the Aircraft safely and one camera operator to manage pan, tilt, focus, and framing.

**Photography Drones**

Photography Drones are highly compact, integrated consumer and prosumer systems. Optimized for single-operator use, they rely on advanced Flight stabilization software, automated shooting modes like tracking or orbiting, and high-resolution, fixed-focal-length camera sensors to capture clean, sharp landscape imagery and commercial photography.

## 4. High-Speed Flight and FPV Systems

This category prioritizes raw mechanical speed, extreme agility, and immediate Pilot feedback over automated safety systems and heavy sensor payloads.

### Racing Drones

Racing quads are built purely for raw speed, structural durability, and power-to-weight performance. They feature minimal carbon fiber frames, high-KV brushless motors, and aerodynamically aggressive propellers. They shed all automated assistance systems like GPS, altitude hold, and collision avoidance to maximize agility, reaching speeds well over 100 mph [160 km/h].

### FPV Drones

First-Person View [FPV] Drones use an onboard camera connected to an ultra-low-latency wireless video transmitter, which beams a live feed directly into a pAir of video goggles worn by the Pilot. This immersive perspective allows Pilots to Fly with precise spatial awareness, navigating tight gaps, diving down buildings, and capturing unique, cinematic action footage.

## 5. Next-Generation Autonomous Implementations

These emerging Technologies use advanced onboard processors and distributed networking to complete complex tasks with minimal human intervention.

### AI Drones

AI-driven Drones carry onboard edge-computing neural networks that process visual data in real time. Instead of relying on a human Pilot or basic pre-programmed GPS coordinates, they use computer vision to navigate dense indoor environments without GPS, identify and track specific target objects, map interior structures on the Fly, and make autonomous safety decisions.

**Swarm Drones**

Swarm Technology involves dozens, hundreds, or thousands of small Drones working together as a single, coordinated system. Rather than being controlled individually, the Drones communicate directly with each other using decentralized algorithms to prevent collisions, share sensory data, and distribute tasks dynamically. They are used for complex military reconnaissance, environmental monitoring, wildfire management, and large-scale synchronized Aerial light shows.

## 6. Niche Environmental Environments

Specialized designs allow operators to collect critical data and navigate spaces that are completely inaccessible to traditional Aerial multi-rotors.

**Underwater Drones**

Operating beneath the water's surface, these systems are categorized as Remotely Operated Vehicles [ROVs] or Autonomous Underwater Vehicles [AUVs]. They trade aerodynamic propellers for waterproof thrusters and use heavy structural hulls to withstand intense deep-sea pressures. Equipped with sonar arrays, underwater lighting, and robotic gripper arms, they inspect subsea pipelines, map marine ecosystems, and assist in marine salvage work.

**Indoor Drones**

Indoor platforms are designed to operate safely inside confined, GPS-denied buildings and industrial spaces. They feature protective structural cages or fully enclosed prop guards to prevent damage during accidental wall impacts. Relying on visual positioning systems [VPS] and ultrasonic altitude sensors, they maintain stable hovers indoors to map underground mines, inspect tight factory ductwork, or secure commercial properties after Hours.

# CHAPTER 8: LEADING DRONE MANUFACTURERS

- **Indian Drone Companies**
- **Global Drone Companies**
- **Drone Startups to Watch**

# Chapter 8 Leading Drone Manufacturers

## Indian Drone Companies

Driven by strong government backing [like the Drone PLI scheme and the *Namo Drone Didi* initiative] and a massive push for self-reliance in defense, India's Drone ecosystem has expanded rapidly.

- **ideaForge Technology:** The undisputed pioneer and market leader in Indian UAVs, holding roughly a 50% market share. They specialize in high-altitude, long-endurance, and Vertical Take-Off and Landing [VTOL] hybrid Drones [such as the SWITCH and NETRA series] heavily utilized by Indian defense forces and industrial sectors for surveillance and mapping.
- **Zen Technologies:** Primarily known for military simulation training, Zen has emerged as a premier player in the high-margin **anti-Drone market**. Their indigenous counter-UAV systems are designed to detect, track, and neutralize rogue Drones.
- **Paras Defence and Space Technologies:** While a broader aerospace and defense player, Paras manufactures high-end agricultural and reconnaissance Drones. They excel in supplying the underlying optical and optronic Technologies [like hyperspectral cameras] that Drone sensors rely on.
- **Droneacharya Aerial Innovations & Drone Destination:** These companies lead the commercial "Drone-as-a-Service" [DaaS] and training sectors. Drone Destination operates as a major Remote Pilot training organization, deploying fleets for massive rural mapping projects [like the SVAMITVA scheme] and agricultural spraying.

## Global Drone Companies

The international arena features a sharp divide between consumer/commercial hardware giants, enterprise AI innovators, and heavy-duty logistics or defense prime contractors.

- **DJI [Da-Jiang Innovations - China]:** The undisputed global juggernaut. Commanding over 70% of the worldwide commercial and consumer Drone market, DJI's Mavic, Matrice, and Agras lines set the gold standard for Aerial photography, industrial inspection, and automated agriculture.
- **Skydio [USA]:** The leading U.S. Drone manufacturer, heavily utilized in enterprise, public safety, and defense sectors. Skydio's core strength is its unparalleled, AI-driven autonomous Flight and 360-degree obstacle avoidance, meaning their Drones can navigate complex, GPS-denied environments without human Piloting errors.
- **Zipline International [USA]:** The global leader in autonomous Drone logistics. Rather than consumer quadcopters, Zipline designs and operates fixed-wing mini-Airplanes that provide instant, on-demand medical and commercial delivery networks across Africa and the U.S.
- **AeroVironment & General Atomics [USA]:** Key defense primes holding monopolies on high-altitude military UAVs. AeroVironment dominates tactical loitering munitions [like the Switchblade series], while General Atomics produces massive, long-endurance platforms like the Predator and Reaper lines.
- **Wingtra [Switzerland]:** A major global player in commercial surveying and mining, famous for its VTOL fixed-wing Drones that combine the vertical takeoff convenience of a quadcopter with the long Flight range of an Airplane.

## Drone Startups to Watch

The newest wave of Drone startups focuses heavily on defense autonomy, stealth configurations, and AI that translates natural language commands into complex multi-Drone missions.

- **Neros Technologies:** A highly funded American defense Tech startup focused on scaling low-cost, mass-manufactured autonomous Drones and unmanned weapon systems tailored for modern high-intensity conflicts.
- **VoltAir:** An ambitious Earth observation startup constructing a globally distributed network of autonomous, weatherized Drones. Using automated charging pads mounted directly onto utility poles, their long-range Drones provide automated, millimeter-scale inspections of power grids and critical infrastructure.
- **Theseus:** Addressing a massive bottleneck in Drone warfare, Theseus builds visual navigation software that allows Drones to operate seamlessly in **GPS-denied or heavily jammed environments**. By matching onboard camera feeds directly to satellite imagery, the system remains completely passive with no radio frequency signature.
- **Ornadyne:** Breaking away from traditional quadcopter designs, Ornadyne engineers high-performance, flapping-wing Drones that mimic the acoustic, visual, and electronic profile of real birds to completely bypass modern counter-UAV defense systems.
- **Degla Inc.:** A software-first startup building an AI framework that translates natural language into multi-Drone mission executions. Instead of manually Piloting or hardcoding coordinates, operators tell the system the high-level objective, and the software coordinates the fleet autonomously.

# PART III

# DRONE REGULATIONS & LEGAL FRAMEWORK

# CHAPTER 9: DRONE RULES IN INDIA

- **Drone Rules 2021**
- **Latest DGCA Updates**
- **NPNT Concept**
- **Digital Sky Platform**

Chapter 9

# Drone Rules in India

As the skies fill with uncrewed systems, maintaining safety, privacy, and national security is critical. India has shifted from a highly restrictive, bureaucratic framework to an agile, digitally automated, and developer-friendly regulatory ecosystem.

Administered by the Ministry of Civil Aviation and the Directorate General of Civil Aviation [DGCA], the current Aviation framework balances rapid Technological growth with strict enforcement.

## 1. Drone Rules 2021: The Foundational Policy

The publication of the **Drone Rules 2021** marked a major turning point for the Indian UAV ecosystem. It replaced the old, complex Unmanned Aircraft System [UAS] Rules with a streamlined framework based on self-certification and trust-based compliance.

**Key Policy Reforms**

- **Reduction in Paperwork:** The number of mandatory regulatory operational forms was cut from 25 to just 5.
- **Abolition of Approvals:** The law removed various security clearances, factory inspections, and Pilot logbook attestations.
- **Expansion of Combined Weight Limits:** The rules cover all operations up to 500 kilograms, easily accommodating large commercial cargo platforms and agricultural sprayers.
- **No-Fly Zones Simplified:** Strict Airspace parameters are clearly divided into automated green, yellow, and red sectors.

## 2. Latest DGCA Updates

The legal landscape continues to evolve to accommodate advanced commercial operations, such as Beyond Visual Line of Sight [BVLOS] Flights and specialized Flight corridors. Under the *Bharatiya Vayuyan Adhiniyam*, the DGCA actively enforces key operational changes.

**Type Certification Mandate**

Every unique Drone design manufactured, sold, or flown for commercial work in India must possess a formal **DGCA Type Certificate**. This ensures that the components, emergency failsafes, and firmware meet national Airworthiness guidelines before entering the field.

**Training Streamlining**

The process for becoming a certified operator is fully structured around **Remote Pilot Training Organisations [RPTOs]**. To Fly any Drone above the Nano weight threshold commercially, a Pilot must hold a valid **Remote Pilot Certificate [RPC]** earned through a designated RPTO.

**Draft Civil Drone Bill**

A new legislative proposal—the *Draft Civil Drone [Promotion and Regulation] Bill*—introduces heavier financial penalties for unauthorized Flights. It also grants law enforcement the direct power to temporarily detain non-compliant hardware on suspicion of safety violations.

## 3. The NPNT (No Permission, No Takeoff) Concept

A key safety mechanism in India's regulatory framework is the **No Permission, No Takeoff [NPNT]** system. This software-driven architecture prevents unauthorized Drone Flights before the Aircraft ever leaves the ground.

**How NPNT Functions**

NPNT is a strict digital handshake between a Drone's onboard Flight controller and the national traffic management framework.

- **The Permission Artifact:** Before beginning a Flight, the operator must request Airspace access via a mobile application or ground control station. If the coordinate path is safe, the system issues a digitally signed token called a **Permission Artifact**.

- **Firmware Validation:** The Pilot uploads this token directly into the Drone. The onboard Flight control firmware validates the digital signature against its internal security keys.
- **The Takeoff Lock:** Without a valid, matching Permission Artifact, the Drone's motors will lock and refuse to arm. If a Pilot tries to bypass this restriction, the Flight logs flag it as a severe compliance breach.

## 4. The Digital Sky Platform

The **Digital Sky Platform** is the single-window online portal that runs India's unmanned traffic management system. Managed by the Airports Authority of India, it integrates registration, Flight clearances, and regional Airspace monitoring into a single digital interface.

### The Three-Tier Airspace Map

Digital Sky maintains an interactive, live-updating geographic map that divides Indian Airspace into three color-coded operational zones:

- **Green Zone:** Airspace extending up to a boundary of 400 feet [120 meters] above ground level. For registered Drones carrying valid credentials, Flight is automated—**no prior Pilot clearance is required** to take off in a green zone.
- **Yellow Zone:** Controlled Airspace that requires active authorization before entry. This includes areas near active Airports or transit corridors. Passing a Flight plan through the system automatically alerts regional Air traffic authorities for a rapid digital sign-off.
- **Red Zone:** Strictly off-limits Airspace. This includes defense installations, international borders, key government buildings, and eco-sensitive national parks. Flights inside a red zone can only proceed with specific, manual clearance from the Central Government.

### Unified Asset Registration

Every Drone operating in India [excluding the ultralight Nano category flown recreationally] must be registered on Digital Sky to receive a **Unique Identification Number [UIN]**. This digital license plate is permanently linked to the operator's verified profile, ensuring clear accountability across the entire national Airspace.

# Chapter 9: Drone Rules in India

- Drone Rules 2021
- Latest DGCA Updates
- NPNT Concept
- Digital Sky Platform

# Chapter 9A Drone Rules in India

India's unmanned aerial vehicle (UAV) ecosystem has shifted away from the restrictive and bureaucratic hurdles of the past. Driven by a vision to make India a global drone hub, the Ministry of Civil Aviation and the Directorate General of Civil Aviation (DGCA) overhauled the regulatory landscape through the landmark Drone Rules 2021. Coupled with continuous platform updates and structural oversight under the newer *Bharatiya Vayuyan Adhiniyam*, the current framework balancing innovation with Airspace safety relies on three pillars: the Drone Rules 2021, the NPNT Concept, and the Digital Sky Platform.

## 1. The Core Framework: Drone Rules 2021

The Drone Rules 2021 completely transformed drone operations in India by replacing a mountain of complex paperwork with a "trust-based" governance model. Security clearances were abolished for domestic operators, import processes were simplified, and unique approvals were reduced from 25 to just a handful.

**Who can Fly commercially?**

To get a Remote Pilot Certificate (RPC) via a DGCA-approved Remote Pilot Training Organization (RPTO), an individual must be an Indian citizen, at least 18 years old, and have passed Class 10 examinations.

## 2. Digital Sky Platform: The Central Command

The Digital Sky Platform acts as India's single-window online portal for all UAV management. Instead of visiting physical ministries, operators handle everything digitally—from basic registration to real-time Flight clearances.

**The Interactive Airspace Map**

Digital Sky features a dynamically updated, color-coded interactive map of Indian Airspace, splitting the country into three distinct Flight zones:

- Green Zone:
- Airspace up to 400 feet (120 meters) AGL (Above Ground Level) that is not designated as a red or yellow zone. For registered drones, no prior Flight permission is required to take off here.
- Yellow Zone:
- Controlled Airspace that historically extended 45 km around Airports, but has been optimized to an 8–12 km radius from the Airport reference point. Flying here requires a pre-Flight plan submission on Digital Sky, with approvals typically processed within 24 hours.
- Red Zone: Strictly no-Fly areas where drone operations are prohibited without explicit, top-tier government authorization. Red zones include international borders, military installations, nuclear facilities, and highly sensitive areas like the Parliament House and Rashtrapati Bhavan in New Delhi.

## 3. The NPNT Concept: "No Permission, No Takeoff"

India's most unique contribution to global UAV traffic safety is the NPNT (No Permission, No Takeoff) architecture. It shifts compliance from a purely legal obligation to a hard firmware-level requirement.The Request: Before any Flight, the Pilot logs into the Digital Sky App or Platform and submits a Flight plan specifying the GPS coordinates, target altitude, and intended time window.

1. The Evaluation: The automated system cross-checks the request against the current interactive Airspace map and any live local restrictions (NOTAMs).
2. The Token: If approved, Digital Sky generates an encrypted digital Permission Token.

3. The Takeoff: This token is uploaded to the drone via its Ground Control Station (GCS). The drone's internal hardware and auto Pilot firmware read the token; if the token is missing, invalid, or expired, the drone's motors will physically refuse to arm.

If a drone drifts outside its geo-fenced permission bubble during Flight, the system automatically flags a protocol violation to ensure compliance.

## 4. Latest DGCA Updates & Safety Compliance

The DGCA regularly refines drone legislation to adapt to rapid market changes, particularly regarding heavy-duty industrial and agricultural systems:

- Strict Type Certification: Manufacturers and importers must obtain a formal Type Certificate for their drone models from the DGCA before those systems can legally be registered or sold in India.
- Battery & Hardware Integrity: For large-scale use cases like agricultural spraying or mapping, replacing a drone battery with an unapproved third-party option instantly voids its Type Certificate and registration. Batteries must comply with safety standards against thermal runaway.
- Enforcement Actions: Flying an unregistered drone above the Nano weight limit, operating in a Red Zone without clearance, or ignoring the NPNT protocol can lead to equipment confiscation, blacklisting from the Digital Sky Platform, and steep penalties ranging from ₹10,000 to over ₹1,000,000 depending on the severity.

Whether you are a casual hobbyist capturing cinematic vacation shots or a commercial Pilot operating heavy agricultural sprayers, staying legal in India means keeping your drone registered, checking your colors on the Digital Sky map, and ensuring your hardware supports NPNT.

# CHAPTER 10: UNDERSTANDING AIRSPACE

- **Green Zone**
- **Yellow Zone**
- **Red Zone**
- **Controlled Airspace**
- **Temporary Restrictions**

Drone.AsiaticInCorp.org

# Understanding Airspace

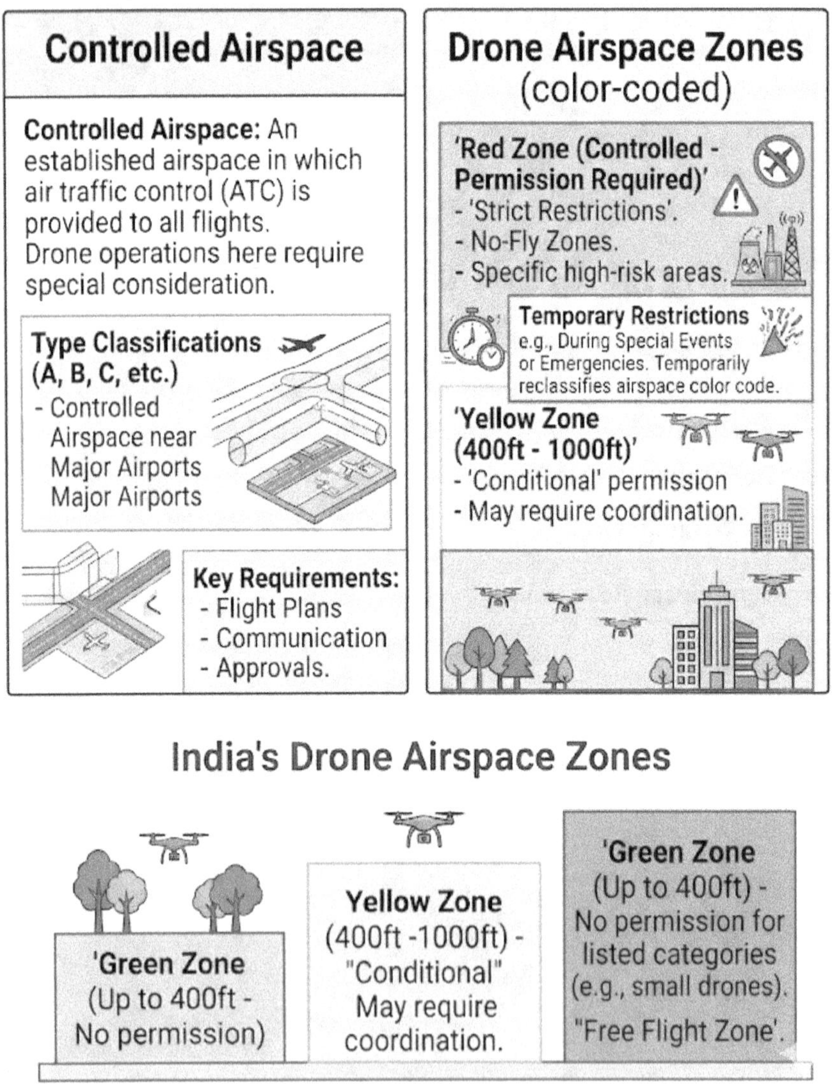

# Chapter 10 Understanding Airspace

Just as roadways are carved into lanes, highways, and restricted zones to keep vehicles moving safely, the sky is partitioned into structured layers. For Unmanned Aerial Vehicle [UAV] Pilots, understanding Airspace layout is just as critical as mastering Flight controls.

In India, the Directorate General of Civil Aviation [DGCA] and the Ministry of Civil Aviation govern these atmospheric layers through a live, interactive map hosted on the Digital Sky Platform. This system relies on strict color-coded zone divisions and specific structural boundaries to integrate Drones into the national Airspace alongside traditional crewed Airplanes and helicopters.

## 1. Green Zone: The Open Skies

Green zones represent low-risk, uncontrolled Airspace where Drone operators can Fly with maximum flexibility and minimal administrative delay.

### Horizontal and Vertical Boundaries

The vertical boundary of a standard green zone extends from ground level up to a hard ceiling of 400 feet [120 meters] Above Ground Level [AGL]. This altitude provides a built-in safety buffer beneath traditional crewed Aircraft, which usually cruise well above 500 feet.

There is also a special buffer zone established around operational Airports. In the area located between a lateral distance of 8 kilometers and 12 kilometers from the outer perimeter of an Airport, the green zone is strictly capped at a lower vertical limit of 200 feet [60 meters] AGL.

### Operational Rules

For any Drone registered on Digital Sky with a valid Unique Identification Number [UIN], Flights within a green zone require **no prior permission or real-time Flight plan approval**. Pilots must simply log their Flight details on the platform. The only major weight exception applies to ultralight Nano Drones [<250 grams] flown purely for recreation; they can operate up to 50 feet [15 meters] AGL in green zones without needing a UIN or Pilot certificate.

## 2. Yellow Zone: Controlled and Monitored Airspace

The yellow zone designates controlled Airspace where Drone Flights present an elevated risk to commercial Aviation, urban structures, or local security, requiring direct oversight before takeoff.

### Triggers for Yellow Airspace

An Airspace is classified as a yellow zone under two clear conditions:

- **Altitude Escalation:** Any Flight path that crosses above the 400-foot [120-meter] ceiling within a designated green zone automatically enters yellow Airspace.
- **Airport Proximity:** The entire Airspace surrounding an active Airport perimeter out to a lateral distance of 8 kilometers is categorized as a yellow zone. Additionally, the space between 8 and 12 kilometers from an Airport becomes a yellow zone for any Flight scaling above 200 feet [60 meters].

### Operational Rules

Flying inside a yellow zone requires **explicit permission from the concerned Air Traffic Control [ATC] authority or Air Defence Clearance center**. Pilots must file a digital Flight plan via the Digital Sky Platform at least 24 Hours in advance. The Drone's firmware must validate this clearance token via the No Permission, No Takeoff [NPNT] protocols before the motors can arm.

## 3. Red Zone: Highly Restricted Airspace

Red zones represent strictly off-limits Airspace where unauthorized Drone operations are entirely prohibited due to significant national security threats, public safety concerns, or environmental sensitivity.

**Typical Red Zone Locations**

- Within a 5-kilometer radius of major international and domestic Airport perimeters.
- Within 25 kilometers of sensitive international land and maritime borders.
- Airspace directly blanketing vital state installations, including military bases, naval ports, nuclear power facilities, and key government buildings like the Parliament house.
- Highly protected eco-sensitive zones, such as prominent national parks and wildlife sanctuaries.

**Operational Rules**

Drones are strictly grounded within a red zone. The only way to legally operate an unmanned Aircraft inside these boundaries is by obtaining manual, specialized clearance directly from the **Central Government** or specific defense ministries. Unauthorized entry into a red zone flags an immediate security violation, leading to heavy regulatory fines, device confiscation, and potential legal prosecution.

## 4. Controlled Airspace: The Shared Aviation Grid

Controlled Airspace refers to generic sectors of the sky where Air traffic control services are actively managed to ensure the safe, orderly flow of all crewed Aircraft.

Because traditional commercial Airliners travel at high speeds and carry human passengers, they always maintain absolute right-of-way over unmanned systems. Drone operations inside controlled Airspace are subject to strict visibility mandates. Pilots must maintain a continuous Visual Line of Sight [VLOS] with their Aircraft during daylight Hours, ensuring they can see and avoid traditional Aircraft entering the immediate vicinity.

## 5. Temporary Restrictions: Dynamic Airspace Shifts

Airspace classifications are never completely permanent. Aviation authorities routinely issue dynamic updates called **NOTAMs [Notices to Air Missions]**, which can temporarily reclassify previously open green zones into highly restricted Airspace.

**Common Triggers for Temporary Adjustments**

- **High-Profile Events:** Major public gatherings, international sports tournaments, or political rallies where large crowds assemble.
- **VIP Movements:** The physical transit of high-ranking state officials or international diplomats.
- **Emergency Incidents:** Active disaster relief zones, active forest fires, or large-scale search and rescue missions where emergency crewed helicopters require exclusive use of local lower altitudes.

Before every deployment, professional Drone Pilots are legally required to cross-verify their intended Flight path coordinates on the Digital Sky map or active NOTAM databases. If a localized safety event has temporarily shifted a green coordinate to red, Pilots must delay their operations until the temporary restriction window officially expires.

- Green Zone
- Yellow Zone
- Red Zone
- Controlled Airspace
- Temporary Restrictions

# CHAPTER 11: DRONE REGISTRATION & COMPLIANCE

- **Registration Process**
- **UIN**
- **Permissions**
- **Documentation**

# Chapter 11 Drone Registration & Compliance

Operating an Unmanned Aerial Vehicle [UAV] in India is no longer an unregulated activity. To ensure accountability, track Air traffic, and uphold national security, the Ministry of Civil Aviation has established a strict legal compliance framework.

Every Drone above a specific weight threshold must be digitally registered, accounted for, and operated under valid permissions. This chapter breaks down the exact steps required to make a Drone legally Airworthy in India, focusing on the registration process, the creation of a Unique Identification Number [UIN], required operational permissions, and essential documentation.

## 1. The Registration Process

All Drone registrations in India are executed digitally through a single-window portal managed by the Directorate General of Civil Aviation [DGCA] called the **Digital Sky Platform**. The platform functions as a centralized database that links hardware, manufacturing data, and owner profiles together.

**Step-by-Step Registration Workflow**

- **Profile Creation:** The individual owner or corporate entity must log onto Digital Sky and create a verified user profile. This step requires Aadhaar verification for individuals or corporate registration data [CIN/GST] for businesses.
- **Drone Category Selection:** The applicant selects whether they are registering a fresh import, a locally manufactured Drone, or an existing legacy Drone that requires normalization.

- **Equipment Mapping:** The operator inputs the Drone's structural characteristics, weight class, and specific serial numbers for critical components, including the Flight controller, chassis, and ground control station.
- **Fee Payment:** A nominal registration fee is processed directly through the portal's integrated payment gateway.

Once the data is submitted, the platform cross-verifies the serial numbers against the manufacturer's uploaded data sheet. If the information matches, the registration is approved automatically.

## 2. Unique Identification Number [UIN]

The **Unique Identification Number [UIN]** is the digital equivalent of a vehicle license plate for Aircraft operating in Indian Airspace.

**Who Needs a UIN?**

According to the Drone Rules, **every Drone operating in India must possess a valid UIN.** The only legal exemption applies to **Nano Drones [weighing under 250 grams],** provided they are flown strictly for non-commercial, recreational purposes within green zones up to a maximum altitude of 50 feet [15 meters]. If a Nano Drone is used for commercial work, it must be registered to obtain a UIN.

**Display and Accountability**

Once a UIN is issued by the Digital Sky Platform, it is permanently tied to that specific Drone frame. The operator must print or engrave this number onto a fireproof identification plate affixed securely to the Drone's chassis. The UIN allows law enforcement and Aviation authorities to scan the Aircraft's wireless broadcast signals or physical frame to instantly identify the registered owner in the event of an Airspace violation or accident.

## 3. Permissions and Approvals

A UIN proves *ownership* of a Drone, but it does not grant an automatic right to Fly anywhere. Operators must secure distinct operational permissions based on the nature of their mission and the location of the Flight.

**Flight Permissions via NPNT**

For any Drone operating outside a recreational Nano profile, Flight access is strictly managed by the **No Permission, No Takeoff [NPNT]** protocol. Before a Flight, the Pilot must log into Digital Sky, select their intended Flight coordinates on the Airspace map, and request a digital permission token.

- If the Flight path falls completely within a **Green Zone**, the permission is granted automatically by the platform within seconds.
- If the Flight extends into a **Yellow Zone**, the request is routed to local Air Traffic Control [ATC] or Air Defence Clearance desks for an active digital sign-off.

**Equipment Import Permissions**

Importing a fully assembled Drone or a Drone kit into India requires a **Directorate General of Foreign Trade [DGFT] Import License,** which is only issued after the Drone model has successfully obtained an equipment type approval from the Ministry of Communications.

## 4. Mandatory Documentation

To stay fully compliant during a field operation, a Remote Pilot must maintain a specific set of active legal documents. If local law enforcement or DGCA inspectors audit a Flight site, failing to produce these documents can lead to immediate equipment confiscation and heavy financial penalties.

**The Compliance Checklist**

- **DGCA Type Certificate:** A copy of the manufacturer's Airworthiness certification proving that the specific Drone model meets national safety standards, featuring mandatory failsafes like "Return-to-Home" [RTH].

- **Remote Pilot Certificate [RPC]:** The active license issued by a DGCA-approved Remote Pilot Training Organisation [RPTO]. This document proves the Pilot has passed the required ground school and Flight proficiency exams.
- **Valid Third-Party Insurance:** Just like motor vehicles, all Drones [excluding recreational Nano units] are legally required to carry active third-party liability insurance to cover property damage or bodily injury caused by an in-Flight mishap.
- **UIN Certificate:** The formal registration sheet downloaded from Digital Sky linking the physical Drone to the operator.
- **Digital Sky Permission Token:** The live, active digital permission artifact generated for that specific day, time, and Flight coordinate boundary.

# CHAPTER 12: PRIVACY, ETHICS & DATA PROTECTION

- **Responsible Flying**
- **Privacy Laws**
- **Ethical Operations**
- **Data Security**

# Chapter 12 Privacy, Ethics & Data Protection

Operating an Unmanned Aerial Vehicle [UAV] safely requires more than just mastering Flight controls or knowing Airspace boundaries. Because Drones carry sophisticated imaging sensors, thermal cameras, and data-logging payloads into the sky, they introduce significant challenges regarding personal privacy, data security, and community ethics.

A professional Drone Pilot must balance Technical capability with social and legal responsibility. This chapter explores how to Fly responsibly, navigate India's modern data and privacy laws, implement ethical operational workflows, and secure Aerial data from digital vulnerabilities.

## 1. Responsible Flying: The Baseline of Trust

Responsible Flying is the practice of prioritizing public safety, psychological comfort, and environmental protection over mission convenience. While Aviation rules focus on keeping hardware from crashing, responsible Flying ensures that the Drone's presence does not disrupt the community below.

**The Core Pillars of Responsible Flight**

- **Mitigating Aerial Nuisance:** Even if a Flight is completely legal within a Green Zone, hovering persistently at low altitudes over residential yards or commercial properties creates noise pollution and public anxiety. Responsible Pilots minimize hover time in sensitive areas and transit quickly to their target coordinates.
- **Community Interface:** Whenever possible, notify local community members or property managers before launching a Drone in a public space. Transparency drastically reduces public friction and unwarranted calls to law enforcement.

- **Environmental & Wildlife Awareness:** Drones can trigger intense stress responses in animals. Flying too close to birds, livestock, or stray animals can cause panicked stampedes or defensive attacks on the Aircraft. Pilots must maintain a respectful distance, especially near nesting sites or forest areas.

## 2. Privacy Laws and the Legal Landscape

In India, Aerial surveillance and data collection are no longer unregulated activities. The legal framework protects citizens from non-consensual tracking and data exploitation from the Air.

### The Right to Privacy as a Fundamental Right

In the landmark *Justice K. S. Puttaswamy v. Union of India* case, the Supreme Court ruled that privacy is a fundamental right under Article 21 of the Indian Constitution. Drone Pilots must recognize that this right extends vertically. Capturing recognizable faces or private residential interiors without explicit consent constitutes a direct violation of this fundamental right.

### The Digital Personal Data Protection [DPDP] Act

India's data handling landscape is governed by the Digital Personal Data Protection [DPDP] Act. Under this framework, any high-definition video, photography, or LiDAR scan that captures identifiable human features constitutes digital personal data.

Drone operators acting as data fiduciaries [those who determine the purpose of data collection] must comply with the DPDP framework. First, under the notice mandate, if a company uses Drones to monitor a specific area, clear notice should be provided to individuals within that space regarding what data is being collected and why. Second, the principle of purpose limitation dictates that data collected for an industrial site inspection cannot be repurposed for employee surveillance or external marketing without fresh consent. Finally, under the right to erasure, individuals have the legal right to request the deletion of their personal data if it was captured incidentally or without explicit authorization.

## 3. Ethical Operations: Navigating Grey Areas

A Drone operation can be perfectly legal under DGCA rules while still being highly unethical. Ethical operations bridge the gap between what a Pilot can legally do and what they should do.

**Best Practices for Ethical Missions**

When conducting real estate photography, an unethical practice is capturing clear views into neighboring backyards or windows. The ethical standard is to frame shots to exclude adjacent properties or to digitally blur unconsenting areas in post-production.

During crowd monitoring or event coverage, Pilots should avoid Flying low over densely populated gatherings without safety nets or notifications. Instead, they should utilize high-altitude zoom lenses from a safe perimeter and post visible entry signs indicating Drone filming is active.

For industrial or site patrols, operators must not use high-definition imagery to track specific employees' casual behavior or breaks. The ethical approach is to restrict data collection strictly to structural assets, security breaches, and safety hazards.

As a golden rule for Aerial imaging, if your Drone captures data that you would consider intrusive if it were filmed over your own home or business, it must be deleted, blurred, or restricted immediately.

## 4. Data Security: Safeguarding the Air-to-Ground Pipeline

Data security involves protecting the digital assets generated by the Drone—from the moment telemetry packets leave the Aircraft to the final storage of raw video on cloud servers. Unsecured Drones can be vulnerable to data interception, unauthorized control, or intellectual property theft.

**Securing the Data Lifecycle**

- **Step 1: Secure the Drone Hardware [Pre-Flight]:** Ensure the physical MicroSD card slots and data ports are locked or tamper-evident. Use high-quality, encrypted storage media on the Aircraft to protect raw files in the event of an in-Flight crash and recovery by unauthorized parties.

- **Step 2: Encrypt the Command and Telemetry Links [In-Flight]:** Utilize Drones featuring AES-128 or AES-256 encryption protocols on the radio links connecting the Ground Control Station [GCS] and the Aircraft. This stops malicious actors from executing man-in-the-middle attacks to hijack the video feed or inject rogue Flight commands.
- **Step 3: Sanitize Ground Control Software [Post-Flight]:** Disable automatic background data-sharing options within the Flight control application. Many consumer Drone apps silently sync Flight logs and low-resolution media files to foreign or third-party cloud servers without the Pilot's deliberate knowledge.
- **Step 4: Implement Secure Storage and Access Control [Data Archiving]:** Store all processed Aerial imagery, mapping data, and thermal sheets on secure servers with restricted user permissions. Apply encryption to data at rest and ensure all transfer paths use secure transport layer protocols like HTTPS or SFTP.

## 5. Compliance Checklist for Mission Planners

Before deploying a commercial Drone surveillance, mapping, or filming mission in India, the mission planner should answer four fundamental safety questions:

- **Consent:** Have we obtained written or explicit permission from the primary land/property owner?
- **Minimization:** Are our camera angles configured to capture only the specific target asset, avoiding accidental capture of surrounding public or private property?
- **Purge Protocol:** Do we have a defined workflow to permanently delete raw files, unneeded footage, and incidental captures once the final client deliverables are generated?

- **Local Laws:**
- Have we cross-checked local police regulations or regional circulars that might place additional, localized restrictions on Aerial filming for that day?

# PART IV

# BECOMING A CERTIFIED DRONE PILOT

# CHAPTER 13: DRONE PILOT ELIGIBILITY

- Age Requirements
- Educational Qualifications
- Medical Fitness
- Required Documents

# Chapter 13 Becoming a Certified Drone Pilot

Navigating the skies with an Unmanned Aerial Vehicle [UAV] is a professional skill that carries regulatory responsibility. In India, Flying any Drone larger than a Nano unit [above 250 grams] for commercial or professional purposes requires a Remote Pilot Certificate [RPC]. This credential is officially granted via the Digital Sky Platform after a candidate successfully completes training at a DGCA-approved Remote Pilot Training Organisation [RPTO].

Before enrolling in ground school and Flight Flight-line training, an aspiring Pilot must meet specific statutory requirements established by the Directorate General of Civil Aviation [DGCA]. This chapter outlines the necessary age limits, academic benchmarks, physical fitness levels, and primary identity documents required to secure an RPC

## 1. Age Requirements

The DGCA has established distinct age limits to ensure that Remote Pilots possess the psychological maturity and physical reflexes required to manage emergency Aerial procedures.

- Minimum Age Threshold: An applicant must be at least 18 years of age to legally enroll in a certified Drone Pilot training course and receive an RPC.
- Maximum Age Limit: The upper operational age limit for a certified Remote Pilot is 65 years. To maintain Airworthiness credentials up to this limit, older Pilots must continuously satisfy ongoing medical and Flight proficiency evaluations.

## 2. Educational Qualifications

Operating commercial Drone hardware involves calculating battery depletion rates, interpreting micro-meteorology data, understanding radio frequency charts, and coordinating with real-time Air Traffic Control [ATC] networks. To ensure Pilots can successfully digest this Technical ground school syllabus, a basic academic baseline is legally mandated.

- Minimum Education: The applicant must have successfully passed their Class 10 [Matriculation] examinations from a recognized educational board.
- Language Proficiency: While advanced Aviation terms are taught during the certification course, a basic understanding of English is highly beneficial, as international Aviation standards, standard Flight control application menus, and DGCA circular updates are primarily issued in English.

## 3. Medical Fitness

While Remote Drone Pilots do not experience the physiological G-forces or altitude changes felt by traditional crewed Aircraft Pilots, they must still maintain excellent situational awareness, quick cognitive reflexes, and dependable motor skills to operate safely from the ground.

- Self-Declaration and Fit Status: Unlike commercial Airline Pilots who require intensive Class 1 or Class 2 institutional medical examinations, a Remote Drone Pilot applicant must generally present a standard Certificate of Medical Fitness.
- The Fitness Evaluation: This document must be signed by a registered medical practitioner [holding an active MBBS degree]. The examination confirms that the candidate does not suffer from any significant physical or mental health conditions—such as severe color blindness, untreated epilepsy, or uncorrected vision impAirments—that could compromise their ability to maintain a continuous visual line of sight [VLOS] with the Aircraft during daytime operations.

## 4. Required Documentation

To streamline the background verification process and link the resulting digital certificate to an individual's official civil profile, candidates must submit a clean set of documents to their chosen training organization before beginning practical simulator modules.

- Proof of Identity and Citizenship: A valid Indian Passport or an Aadhaar Card linked to an active mobile number is mandatory. These documents serve as the primary proof of citizenship and confirm the legal identity of the operator on the Digital Sky network.
- Proof of Age and Education: A Class 10 school leaving certificate, matriculation passing certificate, or official marksheet is required to verify both the applicant's date of birth and their academic eligibility.
- Secondary Identification: Depending on specific RPTO onboarding requirements, candidates may also be asked to provide a secondary government ID, such as a Voter ID card, a permanent account number [PAN] card, or a valid driver's license.
- **Medical Fitness Form:**

  The original physical copy of the medical fitness certificate, formatted to fit DGCA compliance layouts, signed and stamped by a licensed physician.

# CHAPTER 14: DGCA REMOTE PILOT TRAINING ORGANIZATIONS [RPTOs]

- **What is RPTO?**
- **Selecting a Training Institute**
- **Training Fees**
- **Course Duration**

# Chapter 14 DGCA Remote Pilot Training Organizations [RPTOs]

Fulfilling the eligibility criteria is only the first step toward taking your Drone operations to a professional level. To legally operate a Drone in India, your practical skills and theoretical knowledge must be vetted by an authorized body. This validation happens through a specialized network of training hubs setup across the country under direct civil Aviation supervision.

## 1. What is an RPTO?

A **Remote Pilot Training Organisation [RPTO]** is an Aviation training institute explicitly approved and authorized by the Directorate General of Civil Aviation [DGCA] to deliver unmanned Aircraft training.

Think of an RPTO as the Flight school equivalent for unmanned Aviation. Under the current Drone Rules, no private individual, self-taught Flyer, or unaccredited institution has the legal authority to issue a valid Pilot credential. An RPTO acts as the official bridge to the Digital Sky Platform—once you pass their internal assessments, the institute uploads your training record directly to the DGCA database, which automatically generates your official **Remote Pilot Certificate [RPC]**. These organizations are audited regularly by the government to ensure their training Drones, simulators, and Flight fields meet strict national safety standards.

## 2. Selecting a Training Institute

With dozens of approved RPTOs running Flight camps across India, choosing the right training center comes down to your career goals and location. Consider these critical factors before enrolling:

- **Drone Category Accreditation:** Drones are categorized by weight [Micro, Small, Medium, Large]. Most commercial Pilots opt for a **Small Category** certification [2 kg to 25 kg], as it legally covers both smaller Micro builds and heavier industrial mapping or film rigs. Ensure the institute is specifically cleared by the DGCA to train on the weight class you intend to operate.

- **Infrastructure and Airspace:** Look for institutes that feature dedicated, obstacle-free Flying fields located in clear Green Zones. High-quality institutes also provide advanced, state-of-the-art Drone simulators running specialized Flight environments to build muscle memory before you handle real hardware.
- **Industry-Specific Specializations:** Some forward-thinking training hubs offer tailored elective modules alongside the standard base certification. If your goal is to enter specialized fields, look for institutes offering dedicated training tracks in agricultural pesticide spraying, LiDAR geographic mapping, thermal infrastructure inspection, or manual First-Person View [FPV] cinematic Flying.

## 3. Course Duration

Earning your RPC doesn't require months of classroom instruction. The standard DGCA-mandated training curriculum is structured as a short-term, high-intensity boot camp designed to get you safely into the sky within **5 to 7 working days**.

The typical 40-Hour syllabus is cleanly partitioned into sequential phases:

- **Ground School [Days 1 & 2]:** Roughly 14 Hours of classroom or digital instruction covering basic aerodynamics, Aviation meteorology, radio telephony procedures, Flight planning, and the legalities of the Drone Rules.
- **Simulator Practice [Day 3]:** A minimum of several Hours logged on PC-based Flight simulators. This risk-free phase trains your hands to handle pitch, roll, yaw, and throttle adjustments, including orientation recovery drills.
- **Practical Flight Training [Days 4 & 5]:** Hands-on Flight line practice under a master instructor. You will log a mandatory minimum of 4 Hours and 30 minutes of live Airtime, practicing

takeoff/landing sequences, standard figure-eight circuits, basic assembly/maintenance, night-Flying simulations, and critical emergency procedures [like manual overrides during GPS failures].

- **The Final Evaluation:** On the final day, you must clear a 90-minute written theory exam and a live Flight skill assessment judged by an authorized examiner.

## 4. Training Fees

The cost of obtaining a Remote Pilot Certificate varies noticeably depending on the institute's geographic location, its infrastructure scale, and the specific category of Drone you train on.

On average, the standard base fee for a commercial Small/Micro Multi-rotor certification ranges from **₹35,000 to ₹85,000** [plus applicable GST].

**Why the Price Variation?**

Institutes in premium metro zones or those tied to established aerospace academies often charge toward the higher end [upwards of ₹75,000 to ₹1,00,000] because they bundle premium amenities into the package, such as on-campus hostel accommodation, meals, job placement assistance, and access to enterprise-grade industrial hardware. Conversely, specialized or regional agricultural Drone training programs supported by state grants or agricultural universities can often be accessed closer to the lower baseline.

While the upfront cost is a noticeable investment, an active commercial Pilot can often break even within their first few months of professional field operations.

## 5. Summary Matrix for Aspiring Pilots

| Feature | Standard Requirement / Expectation | |
|---|---|---|
| **Course Length** | 5 to 7 Days [approx. 40 Hours total] | |
| **Average Cost** | ₹35,000 to ₹85,000 + GST | |

| Feature | Standard Requirement / Expectation | |
|---|---|---|
| **Minimum Airtime** | 4.5 Hours of live Flight logs | |
| **Exam Pass Mark** | 70% minimum score on the theory evaluation | |
| **Certificate Validity** | 10 Years [Fully verifiable on the Digital Sky portal] | |

# Chapter 15: DGCA Drone Pilot Certification Process

- Step-by-Step Guide
- Enrollment
- Training
- Assessment
- Certification

# Chapter 15 DGCA Drone Pilot Certification Process

Meeting the basic eligibility criteria and picking a training institute sets your journey in motion. However, actually obtaining your license requires navigating a standardized path overseen by Aviation authorities.

In India, the transition from an aspiring student to a licensed Flyer follows a formalized, multi-step pipeline. This framework connects your physical training with the digital infrastructure of the **eGCA portal** and the **Digital Sky Platform**, ensuring that every commercial Pilot in the sky is thoroughly vetted and fully traceable.

## 1. Step 1: Portal Enrollment & Onboarding

Before you handle a real Drone or step into a classroom, your digital Aviation profile must be formally established with the government.

**Creating Your eGCA ID**

While the Digital Sky Platform remains the live hub for Airspace maps and real-time Flight plan approvals, individual Pilot onboarding and certificate management are handled through the **eGCA portal**. You must register on the portal specifically under the "Remote Pilot" category. This process uses Aadhaar-linked mobile verification to tie your legal identity to your future Pilot profile.

**Document Submission at the RPTO**

Once your eGCA profile is live, you must submit your physical documentation to your chosen DGCA-approved Remote Pilot Training Organisation [RPTO]. The required package includes your Class 10 marksheet [verifying age and education], a valid government photo ID, passport-size photographs, and your original Medical Fitness Certificate signed by a registered MBBS doctor.

## 2. Step 2: Ground School & Practical Training

With your registration approved, your formal training begins. The RPTO takes you through a high-intensity, structured curriculum that balances classroom theory with practical cockpit time over 5 to 7 days.

### Ground School [Theory Modules]

The initial phase consists of roughly 14 classroom Hours covering the baseline knowledge required to navigate shared civilian Airspace safely. Key subjects include:

- **Air Law and Regulations:** Understanding the Drone Rules, your legal liability as a Pilot, and Airspace classifications.
- **Flight Aerodynamics:** How multi-rotors generate lift, maintain stability, and combat battery depletion.
- **Meteorology:** Reading wind shear tables, anticipating turbulence, and understanding how Air density affects Flight performance.
- **Emergency Management:** Formulating response protocols for unexpected battery fires, signal drops, or mid-Air system failures.

### Simulator Training

Before taking off with actual hardware worth lakhs, you must log specialized Hours on an institutional Flight simulator. This risk-free environment trains your fingers to control pitch, roll, yaw, and throttle. You will practice maintaining a stable hover, recovering from sudden spatial disorientation, and Flying without GPS assistance [Attitude Mode].

### Live Flight Practice

Once your simulator performance is validated, you move onto the live Airfield. Under the direct supervision of a certified instructor, you must log a mandatory minimum of 4 Hours and 30 minutes of actual Airtime. This phase moves you through takeoff/landing drills, precise square circuits, figure-eight maneuvers, night-Flying simulations, and manual emergency overrides.

## 3. Step 3: Evaluations and Assessments

Completing your training Hours does not guarantee an automatic license. You must demonstrate clear competence by passing a dual-part final evaluation conducted at the training facility.

RPTO Training Completed

↦

Theory Examination [Online via Portal] ⟶

Requires ≥ 70% to Pass

⟶ Practical Skill Check [Field Flight] ⟶ Validated by Authorized Examiner

**The Theory Examination**

You must clear a comprehensive written theory exam hosted directly on the online testing portal. The test evaluates your recall of Airspace laws, weather patterns, and emergency checklists. The passing threshold is strictly set at a minimum score of 70%, and applicants are permitted up to 3 separate attempts to clear this bottleneck.

**The Practical Skill Assessment**

The practical check puts you alone on the Flight line with an authorized Drone examiner. You are required to perform a complete pre-Flight safety inspection, execute a series of precise Flight maneuvers smoothly, demonstrate clean orientation control, and safely land the Aircraft following a simulated system failure.

## 4. Step 4: Certificate Issuance

Once you clear both the written and practical hurdles, the administrative process transitions back to the digital portals for final verification.

## RPTO Application Lodging

The training institute acts as your verification sponsor. Within 7 days of your successful final field assessment, the RPTO uploads your authenticated Flight logs, exam scores, and practical evaluation metrics directly onto the government Aviation database, initiating the formal application for your license.

## Receiving Your Remote Pilot Certificate [RPC]

Aviation authorities review the uploaded records to ensure compliance with national standards. Once validated, your official **Remote Pilot Certificate [RPC]** is digitally generated and issued via the platform. This processing window typically takes between 7 to 15 working days.

The resulting digital certificate features a secure, verifiable QR code and can be printed or stored on your mobile device. Under current regulations, a standard commercial RPC remains completely valid for 10 years from the date of issuance, serving as your official passport to a professional career in the skies.

# Chapter 16: Drone Pilot License & Career Roadmap

- **Certification Validity**
- **Renewals**
- **Upgrading Skills**
- **Career Progression**

# Chapter 16 Drone Pilot License & Career Roadmap

Securing your Remote Pilot Certificate [RPC] marks the transition from an intermediate Flyer to a certified professional. However, obtaining the physical certificate is not the final destination—it is the starting block.

The unmanned Aviation sector is dynamic, with rapidly evolving hardware capabilities, shifting regulatory landscapes, and expanding industrial use cases. To turn your certificate into a long-term, high-yielding career, you must understand the rules governing certificate lifecycle management and map out a strategic professional roadmap.

## 1. Certification Validity and Lifespan

Once issued by a DGCA-approved training organization and logged onto the eGCA portal, a standard Remote Pilot Certificate features a highly stable regulatory lifespan.

- **10-Year Operational Window:** Your digital RPC remains fully legal and valid for a continuous period of **10 years** from the formal date of issuance.
- **National Recognition:** The certificate functions as a single, standardized federal credential. It is legally recognized across all Indian states and union territories, allowing you to deploy on cross-border industrial operations without needing regional re-validation.
- **Age Ceiling:** While the certificate itself lasts a decade, your authorization to act as a Remote Pilot faces a hard stop when you reach **65 years of age**, which is the maximum legal age limit for Drone operators under current Indian civil Aviation rules.

## 2. The Renewal Process

As your 10-year validity window approaches its expiration date—or if your operational continuity lapses—you must initiate a formal renewal pipeline to keep your credentials active.

Rather than forcing experienced Pilots to repeat the entire week-long novice training curriculum, the DGCA structures renewals as a streamlined proficiency checkpoint:

- **The Refresher Module:** Pilots must enroll in a short, condensed refresher course at an authorized Remote Pilot Training Organisation [RPTO]. This session focuses heavily on updated Drone Rules, newly implemented Airspace protocols, and modern UTM [UAS Traffic Management] changes.
- **Medical Re-Verification:** You must secure a fresh Certificate of Medical Fitness from a registered MBBS doctor, proving that your physical reflexes, vision, and cognitive faculties still meet baseline safety demands.
- **The Proficiency Practical:** An authorized instructor executes a brief field skill check to verify that your Flight handling, emergency management, and pre-Flight evaluation habits remain sharp.

Once the RPTO validates your refresher metrics, your renewal token is logged on the Digital Sky Platform, extending your operational clearance for an additional block of time.

## 3. Upgrading Skills: Moving Beyond the Basics

A standard entry-level RPC typically clears you to operate a specific category of Aircraft—usually a **Small Category Multi-rotor** [Drones weighing between 2 kg and 25 kg]. While this is the most flexible baseline for commercial work, staying restricted to basic daylight photography limits your earning potential. To maximize your value, consider these three professional upgrade tracks:

### Weight Class Escalations

If you want to transition into heavy-lift logistics, long-range cargo delivery, or heavy industrial operations, you must return to an RPTO to upgrade your license class from Small to **Medium Category** [25 kg to 150 kg] or **Large Category** [above 150 kg]. This requires logging additional training Hours on larger frames and clearing weight-specific structural exams.

**Alternate Design Endorsements**

Multi-rotors are excellent for precision hovering, but they lack the battery efficiency required to cover massive stretches of land. Upgrading your skill set involves securing endorsements for alternate Airframes:

- **Fixed-Wing Drones:** Shaped like traditional Airplanes, these are highly sought after for large-scale linear mapping, pipeline inspections, and corridor monitoring.
- **Hybrid VTOL [Vertical Take-Off and Landing]:** These systems combine the vertical take-off convenience of a multi-rotor with the high-speed cruising efficiency of a fixed-wing craft.

**Advanced Operational Skill Sets**

True specialization comes from mastering advanced payloads and complex Flight profiles:

- **BVLOS [Beyond Visual Line of Sight]:** Training to operate Drones safely over long distances where the Aircraft is no longer visible to the naked eye. This skill is critical for upcoming e-commerce logistics and widespread agricultural tracking.
- **Data Processing Specializations:** Learning to translate raw Drone outputs into actionable industrial intelligence, such as processing LiDAR point clouds, generating 3D orthomosaic maps, or interpreting thermal imagery for solar farm audits.

## 4. Career Progression: The Drone Industry Roadmap

The Drone industry has evolved into a multi-tiered job market. Certified Pilots can chart their professional growth across four primary career trajectories:

**Field Operations Track**

- **Junior Drone Pilot:** Conducting routine, localized visual-line-of-sight operations such as basic real estate shoots, wedding cinematography, or entry-level construction progress photography.

- **Senior Enterprise Pilot:**
- Leading complex, high-risk industrial deployments. This includes managing multi-million rupee asset inspections, navigating hazardous thermal environments inside industrial plants, or running large-scale crop-spraying programs.
- **Flight Operations Manager:** Transitioning away from the sticks to supervise whole corporate Drone fleets. You will manage mission compliance checklists, schedule equipment maintenance logs, secure complex ATC clearances, and oversee field safety protocols.

**Training & Academia Track**

- **Certified Flight Instructor:** After accumulating significant field experience and clearing a specialized "Train the Trainer" program, you can join an approved RPTO as a Flight line instructor, shaping the next generation of incoming student Pilots.
- **Chief Remote Pilot Instructor [CRPI]:** Overseeing the entire educational layout of a training institute, certifying internal exam scores, and managing regulatory audits conducted by DGCA inspectors.

**Data Analytics Track**

- **Drone GIS Analyst / Photogrammetrist:** Using powerful mapping engines to convert raw Aerial imagery into highly detailed geographic data maps, volumetric asset calculations, or topological contours for infrastructure developers and government planning bodies.

## 5. Summary Checklist for Career Launch

Before booking your first commercial contract, ensure your operational workflow matches this professional sequence:

- **Verify Your Logs:** Maintain a neat, continuous digital record of all Flight times, equipment configurations, and coordinates inside a professional Pilot logbook.

- **Keep Insurance Active:** Confirm that your third-party liability Drone insurance policy is continuously updated to shield your business from unexpected financial claims during field accidents.
- **Build an Industry Portfolio:** Group your Flight accomplishments by industrial niche [e.g., Agriculture, Infrastructure, Cinematography] to pitch effectively to corporate clients.
- **Monitor Digital Sky:** Log onto the Digital Sky platform regularly to scan for real-time circular updates, keeping your field operations ahead of any sudden legal changes.

# PART V

# Drone Pilot

# CHAPTER 17: AIR REGULATIONS

- DGCA Regulations
- Air Law
- Aviation Regulations

# Chapter 17 Air Regulations

Entering the cockpit of an unmanned Aircraft requires a foundational transition from a casual hobbyist mindset to that of a professional Remote Pilot. Because Drones share the lower boundaries of the sky with helicopters, commercial Airliners, and military transport Aircraft, they are legally bound by national and international Aviation frameworks.

The Air Regulations module forms the core of the DGCA ground school syllabus. It equips trainees with the mandatory knowledge of civil Aviation law, statutory definitions, operational rules, and the legal liabilities required to clear the Remote Pilot theoretical examination.

## 1. Statutory Framework and Air Law

Air law is not a singular rulebook, but a layered system of international treaties and national legislations designed to keep global Air travel orderly and safe.

The foundational bedrock of all modern Air law is the Convention on International Civil Aviation, signed in Chicago in 1944. Administered by the International Civil Aviation Organization [ICAO], this treaty sets global standards for Airspace management, Aircraft registration, safety audits, and accident investigations. Under ICAO guidelines, every sovereign nation maintains absolute and exclusive authority over the Airspace directly above its territory.

India exercises its Airspace sovereignty through two core pieces of parliamentary legislation, which grant civil Aviation authorities the legal power to govern the skies. First, the Aircraft Act of 1934 serves as the primary statute giving the Central Government the authority to control the manufacture, possession, use, operation, import, and export of all Aircraft—explicitly including unmanned Aircraft systems—within

India. Second, the Aircraft Rules of 1937 establish the broader regulatory framework implementing the provisions of the 1934 Act. This document details the licensing of Flight crew, Airworthiness of frames, Air traffic flow protocols, and enforcement mechanics for safety violations.

## 2. DGCA and the Ministry of Civil Aviation

To understand how Drone rules are updated and enforced, a Pilot must map the administrative hierarchy that governs civil Aviation in India.

The Ministry of Civil Aviation [MoCA] is the apex government ministry responsible for formulating national policies, drafting major Aviation bills, and driving the strategic growth of India's civil Flight sector.

The Directorate General of Civil Aviation [DGCA] is the statutory regulatory body functioning under MoCA. Headquartered in New Delhi, the DGCA is the practical enforcement arm of Indian Aviation. Its core responsibilities include the registration and tracking of civil Aircraft, the licensing of Pilots, Aircraft maintenance engineers, and Air traffic controllers, and the formulation of Civil Aviation Requirements [CARs] and training circulars. It also handles issuing authorizations to Remote Pilot Training Organisations [RPTOs] and investigating Drone mishaps, dangerous Airspace incursions, and regulatory non-compliance.

The operational pipeline flows directly from MoCA down to the DGCA, which then manages and enforces the rules through the automated Digital Sky Platform used by Remote Pilots.

## 3. The Drone Rules: India's Unmanned Legal Code

While traditional Aircraft must follow thousands of pages of the historic Aircraft Rules, the explosive growth of unmanned Aviation led the Ministry of Civil Aviation to carve out a highly specific, simplified code known as the Drone Rules. This framework replaces lengthy manual approvals with a modern, automated system.

Key legal tenets of the Drone Rules include:

- **The UIN Mandate:** Operating an unregistered Drone [except for a Nano category Drone flown purely for recreation] is a direct federal offense. The Unique Identification Number [UIN] acts as a digital license plate that must be anchored to a fireproof plate on the Aircraft chassis.
- **The Absolute Ceiling:** The maximum legal Flying altitude for a Drone operating in standard uncontrolled Airspace is strictly capped at 400 feet [120 meters] Above Ground Level [AGL]. Scaling past this boundary pushes the Drone into commercial Flight lanes, requiring explicit ATC authorization.
- **Visual Line of Sight [VLOS]:** Remote Pilots are legally mandated to maintain a continuous, direct, unaided visual connection with their Drone throughout the entire Flight profile. Utilizing secondary aids like binoculars to track a distant Drone is illegal. FPV goggles are only permitted if a designated visual observer stands next to the Pilot to monitor the ambient Airspace.
- **Commercial Third-Party Insurance:** Just like road vehicles, no Drone [excluding recreational Nano units] can turn its props on a commercial mission without an active third-party liability insurance policy to safeguard property and citizens below.

## 4. Operational Classification of Drones

The DGCA categorizes all Unmanned Aircraft Systems based on their Maximum Take-Off Weight [MTOW], which includes the dead weight of the battery, structural payload, cameras, and additional sensory attachments. This weight determines the exact level of regulatory oversight required.

- **Nano Category:** Drones weighing up to 250 grams. They represent the lowest safety risk. They do not require a UIN, type certificate, or Pilot certificate when flown recreationally within green zones up to 50 feet AGL.

- **Micro Category:** Drones weighing between 250 grams and 2 kilograms. These are commonly used for wedding cinematography, real estate, and basic inspection work. They require full digital registration and an RPC if operated commercially.
- **Small Category:** Drones weighing between 2 kilograms and 25 kilograms. This is the workhorse bracket for enterprise deployment, covering high-end filmmaking platforms, thermal mapping platforms, and localized agricultural spraying systems. An RPC is mandatory for all Flights within this weight bracket.
- **Medium Category:** Drones weighing between 25 kilograms and 150 kilograms. Typically used for large-scale industrial cargo movement, multi-hectare agricultural spraying, or specialized scientific research.
- **Large Category:** Drones weighing above 150 kilograms. These function as heavy-duty industrial Aircraft and are subject to stringent Airworthiness evaluations akin to traditional crewed Aircraft.

## 5. Penalties for Air Law Non-Compliance

To maintain discipline in shared Airspace, the civil Aviation framework outlines severe legal penalties for Pilots who compromise public safety or compromise restricted areas.

Operating a Drone without a valid UIN, an active Remote Pilot Certificate, or necessary insurance can lead to equipment seizure and direct regulatory fines.

Deliberately steering a Drone into a designated Red Zone—such as an active Airport runway path or near a sensitive border establishment—is classified as a major security violation. This triggers criminal prosecution under Indian penal laws alongside immediate cancellation of the operator's Pilot credentials.

# CHAPTER 18: PRINCIPLES OF FLIGHT

- **Aerodynamics**
- **Lift**
- **Weight**
- **Drag**
- **Thrust**
- **Stability**

# Chapter 18 Principles of Flight

The movement of an Unmanned Aerial Vehicle [UAV] through the atmosphere is governed by the laws of physics. Whether operating a heavy fixed-wing surveying platform or a compact multi-rotor quadcopter, an Aircraft must continuously interact with the surrounding Air to achieve controlled Flight.

The Principles of Flight module covers fluid dynamics and aerodynamic forces. Understanding these concepts allows a Remote Pilot to anticipate how changes in speed, payload, and environmental conditions alter Aircraft performance.

## 1. Fundamentals of Aerodynamics

Aerodynamics is the study of how gases interact with moving bodies. Because Air is a physical fluid with mass and density, it exerts continuous pressure on any object moving through it.

**Bernoulli's Principle**

A key pillar of aerodynamic lift is Bernoulli's Principle, which states that as the velocity of a moving fluid [or gas] increases, the static pressure within that fluid simultaneously decreases.

In fixed-wing Aircraft and traditional helicopter rotors, this principle is applied using an Airfoil—a structure with a curved upper surface and a flatter lower surface. As forward propulsion forces Air across the Airfoil, the Air traveling over the curved top surface must move faster than the Air passing underneath. This speed differential creates a low-pressure pocket on top of the blade and a high-pressure zone beneath it, pushing the wing upward.

### Newton's Third Law of Motion

While Bernoulli's principle explains pressure differentials, lift is also generated by physical deflection. Newton's Third Law states that for every action, there is an equal and opposite reaction. When a Drone's spinning propeller blades tilt at an angle, they physically push mass volumes of Air downward [the action]. The opposite reaction is an upward force exerted by the Air against the propeller blades, lifting the Drone into the sky.

## 2. The Four Forces of Flight

Every Aircraft in Flight is constantly acted upon by a balance of four opposing aerodynamic forces: Lift, Weight, Thrust, and Drag. A Remote Pilot controls the Aircraft by adjusting the relationship between these forces.

[Image diagram showing the four forces of Flight acting on an Aircraft]

### Lift

Lift is the upward aerodynamic force generated by wings or spinning rotor blades that overcomes the pull of gravity. In a multi-rotor Drone, lift is regulated purely by changing the revolutions per minute [RPM] of the brushless motors. To ascend, the motors must spin fast enough to generate a total lifting force that exceeds the physical mass of the Aircraft.

### Weight

Weight is the downward force pulling the Aircraft toward the center of the earth due to gravity. The total weight of a Drone is not just the empty weight of its frame; it includes the battery, internal Flight computers, structural accessories, and any attached payloads like cameras or thermal sensors. Weight acts through a single theoretical point known as the Center of Gravity [CG]. Managing weight is critical, as an overloaded Drone will rapidly drain its batteries and struggle to generate sufficient lift during emergency climbs.

### Thrust

Thrust is the forward force that moves an Aircraft through the Air, overcoming aerodynamic resistance. In a fixed-wing Drone, thrust is produced by a dedicated forward-facing or rear-facing propeller. In a multi-rotor system, thrust and lift are combined. To move forward, a quadcopter changes the relative speed of its rotors to tilt its entire frame forward. This directs a portion of the downward lifting thrust backward, pushing the Drone horizontally.

### Drag

Drag is the rearward retarding force caused by disruption of Airflow around the Drone's structure. It acts as aerodynamic friction, directly opposing forward thrust. Drag increases exponentially as Air velocity climbs. It is divided into two primary types:

- **Parasitic Drag:** The resistance created by non-lifting structural components pushing through the Air, such as landing gear legs, exposed wiring, square battery compartments, and camera gimbals.
- **Induced Drag:** The natural aerodynamic byproduct of generating lift. As propeller blades create pressure differentials, Air wraps around the blade tips, creating performance-draining vortices that hold the Aircraft back.

When an Aircraft is in a steady, unaccelerated straight-and-level hover, Lift equals Weight, and Thrust equals Drag. Shifting this equilibrium causes the Drone to climb, descend, accelerate, or stop.

## 3. Aircraft Stability and Control Axes

Stability refers to an Aircraft's ability to maintain its intended Flight path and spontaneously return to a steady state after being disrupted by external forces, such as sudden wind gusts or turbulent updrafts.

### The Three Axes of Flight

An Aircraft maneuvers through three-dimensional space by rotating around three mutually perpendicular axes that intersect precisely at the Drone's Center of Gravity.

[Image showing the longitudinal lateral and vertical axes intersecting at an Aircraft's center of gravity]

- **The Longitudinal Axis [Roll]:** An imaginary line running straight from the front nose of the Drone to the rear tail. Rotation around this axis is called Roll. It causes the Drone to tilt sideways, shifting the Aircraft left or right.

- **The Lateral Axis [Pitch]:** An imaginary line running horizontally from wingtip to wingtip or from the left rotors to the right rotors. Rotation around this axis is called Pitch. Tilting the nose downward or upward causes the Drone to slide forward or backward.
- **The Vertical Axis [Yaw]:** An imaginary line running straight down through the center of the Drone from top to bottom. Rotation around this axis is called Yaw. It changes the directional heading of the Drone's nose, rotating it clockwise or counterclockwise.

**Managing Stability in Multi-Rotors**

Traditional Airplanes use movable mechanical Flight surfaces like ailerons, elevators, and rudders to control these axes. Multi-rotor Drones handle stability digitally through an onboard Flight Controller equipped with an Inertial Measurement Unit [IMU].

The IMU uses micro-gyroscopes and accelerometers to sense unexpected shifts in roll, pitch, or yaw hundreds of times per second. If a wind gust suddenly tilts the Drone's nose upward [a pitch disruption], the Flight controller instantly accelerates the rear motors and slows down the front motors. This micro-adjustment balances the forces and returns the platform to a stable, level position without requiring manual Pilot input.

# Chapter 19: Airspace & Navigation

- **Navigation Basics**
- **GPS**
- **Maps**
- **Charts**
- **Flight Planning**

# Chapter 19 Airspace & Navigation

Safely operating an unmanned Aircraft requires a clear understanding of the Airspace above us. Unlike driving on a flat road, Flying a Drone requires sharing a complex three-dimensional environment with commercial Airliners, military Aircraft, and emergency helicopters.

This module breaks down how the sky is mapped, how your Aircraft figures out where it is, and how you must plan your Flights to stay entirely within legal and safe boundaries.

## 1. Airspace Navigation Basics

To Pilot an Aircraft Remotely, you must look at the world through the eyes of an Aviator. This means understanding how positions are mapped globally and how altitude is calculated.

- **Latitude and Longitude:** This global grid allows you to pinpoint any location on Earth. **Latitude** lines run horizontally [measuring north or south of the Equator], while **Longitude** lines run vertically [measuring east or west of the Prime Meridian]. In Drone Flight logs, these coordinates are displayed in Decimal Degrees
- **Altitude vs. Elevation vs. Height:** Knowing your distance from the ground prevents catastrophic mid-Air collisions.
- **Elevation:** The vertical distance of a fixed point on the ground above Mean Sea Level [MSL].
- **Altitude [MSL]:** The vertical distance of your Drone Flying above Mean Sea Level.
- **Height [AGL]: Above Ground Level.** This is the most critical metric for Drone Pilots. If you are standing on a hill that is 500 feet

high, and your Drone flies 100 feet above you, your altitude is 600 feet MSL, but your height is 100 feet AGL.

## 2. How GPS Works in Drone Operations

The Global Positioning System [GPS] is the invisible spine of modern uncrewed Aviation. Without a solid satellite lock, a Drone cannot hover statically in place against the wind, execute automated Flight paths, or safely trigger its Return-to-Home [RTH] sequence.

**The Trilateration Process**

Your Drone does not "transmit" anything to space; it is a passive listener. GPS satellites orbiting Earth broadcast incredibly precise time-stamped signals. The Drone's internal GPS receiver picks up these signals and calculates exactly how long they took to travel from space to your Aircraft at the speed of light.

To find your exact 3D position in the sky, the Drone uses a mathematical process called **trilateration**:

- **3 Satellites:** Narrows your position down to two intersecting points on Earth [giving you latitude and longitude].
- **4 Satellites:** Pinpoints your exact location and provides your **altitude**.
- **7 to 12+ Satellites:** This is the standard minimum required for safe Drone operations. The more satellite locks your Drone maintains, the smaller its margin of error becomes, preventing "GPS drift."

**Warning: The "Toilet Bowl" Effect**

If your Drone's internal compass or GPS experiences electromagnetic interference [from large metal structures or concrete rebar], it can become confused. The Drone may try to correct its position based on conflicting data, causing it to Fly in rapidly expanding, uncontrollable circles. If this occurs, switching to manual ATTI [Attitude] Mode allows the Pilot to steer without relying on GPS.

## 3. Aviation Maps, Charts, and Airspace Zones

Standard road maps are practically useless for Flight planning. Pilots rely on sectional aeronautical charts and digital Airspace maps to identify terrain hazards, Flight restricted zones, and tower structures.

**The Indian Airspace Zoning System**

Under the DGCA [Directorate General of Civil Aviation] framework, the Airspace across India is divided into three live, color-coded zones. You must consult the interactive map on the official **Digital Sky Platform** before every takeoff.

as

| Airspace Zone | Flight Restrictions & Permissions |
|---|---|
| **Green Zone** | **Free Flight Zone.** Airspace up to **400 feet [120 meters] AGL**. No prior Flying permission is required for registered Drones, though you must still log your Flight. It also extends up to 200 feet AGL in the area between 8 to 12 km from an active Airport perimeter. |
| **Yellow Zone** | **Controlled Airspace.** This includes all Airspace above 400 feet in a green zone, or any Airspace located closer to an Airport [within 8 to 12 km under 200 feet, and completely between 5 to 8 km from an Airport perimeter]. **Prior permission from Air Traffic Control** [ATC] **is mandatory** via Digital Sky. |
| **Red Zone** | **Strict No-Fly Zone.** Includes areas within a 5 km radius of operational Airports, international borders, military installations, vital government buildings [like Parliament], and national parks. Flying here is strictly prohibited unless explicitly authorized by the Central Government. |

## 4. Pre-Flight Planning and Digital Sky Compliance

Flight planning is the structured process of mitigating risk before your Drone leaves the ground. Operating legally requires a blend of digital compliance and localized site assessment.

## Step-by-Step Mission Planning Sequence

**1.Check the Digital Sky Map:**Step 1: Zone Verification.

Log into the Digital Sky platform and cross-reference your intended GPS coordinates against the interactive Airspace map to ensure you are not infringing on a Yellow or Red zone.

**2.Confirm UIN and Insurance:**Step 2: Documentation Check.

Ensure your Drone has its Unique Identification Number [UIN] physically marked on the frame. Verify that your mandatory third-party liability insurance policy is active to protect against property damage or personal injury.

**3.Check NOTAMs and Weather:**Step 3: Environmental Assessment.

Review local NOTAMs [Notices to Airmen] for temporary Flight restrictions [e.g., VIP movements]. Check wind speeds, visibility conditions, and K-index values [solar activity that can disrupt GPS signals].

**4.Secure Takeoff Permission:**Step 4: Obtain Digital Clearance.

Submit your Flight plan. If Flying an NPNT [No Permission, No Takeoff] compliant Drone, download the encrypted permission artifact onto your Drone's ground station hardware. The Drone's firmware will physically refuse to spin its motors without this digital green light.

**5.Conduct Physical Site Survey:**Step 5: Pre-Flight Checklist.

Establish a clear, unobstructed take-off and landing zone. Scan the immediate area for overhead power lines, tall trees, or clusters of people. Ensure you can maintain an absolute **Visual Line of Sight [VLOS]** with the Aircraft throughout the Flight.

# Chapter 20: Meteorology

- **Weather Systems**
- **Wind Analysis**
- **Visibility**
- **Rain & Storms**
- **Weather Forecasting**

# Chapter 20 Aviation Meteorology Guide for Drone Pilots

Mastering Aviation meteorology is fundamental to safe Flight execution because atmospheric conditions directly govern an Aircraft's performance, stability, and overall Airworthiness. For Unmanned Aerial Vehicle [UAV] operators, checking the weather isn't just about avoiding a drizzle—it is a critical safety and regulatory requirement. This guide breaks down the core concepts of Drone meteorology, including weather systems, wind metrics, visibility limitations, storm lifecycles, and Aviation forecasting.

## 1. Understanding Aviation Weather Systems

Weather systems are driven primarily by the unequal heating of the Earth's surface, which creates distinct Air masses characterized by uniform temperature and moisture levels. When two different Air masses collide, they form a boundary called a front.

- **Cold Fronts:** A cold front occurs when a dense, cold Air mass wedges beneath a warm Air mass, rapidly forcing the warm, moist Air upward. This rapid displacement triggers turbulent weather, sudden temperature drops, and heavy convective cloud formations like cumulonimbus clouds.
- **Warm Fronts:** A warm front involves a warmer Air mass overriding a retreating colder Air mass. This leads to a gradual lifting process that produces widespread, layered stratus clouds and steady, prolonged precipitation.

### The Indian Monsoon Context

In India, regional weather systems play a decisive role in Flight planning. The massive seasonal shifts of the Southwest and Northeast monsoon systems introduce prolonged periods of heavy rainfall, low cloud bases, and widespread wind shifts. Additionally, winter months frequently bring western disturbances that disrupt Drone operations across northern territories with sudden atmospheric changes.

## 2. Advanced Wind Analysis and Localized Hazards

Wind analysis is critical for navigation, battery management, and maintaining aerodynamic control during take-off and landing phases. Wind is simply the movement of Air traveling from areas of high pressure to areas of low pressure—a process influenced on a global scale by the Coriolis force, which deflects moving Air due to the Earth's rotation.

### Local Thermal and Terrain Microclimates

On a localized level, Pilots must constantly account for thermal variations that generate diurnal wind patterns:

- **Sea and Land Breezes:** A sea breeze occurs during the day when the land heats up faster than the water, drawing cooler Air inland. A land breeze occurs at night as the land cools down rapidly, reversing the flow.
- **Mountain Air Movements:** Mountainous terrain introduces valley breezes [warm Air ascending slopes during the daytime] and katabatic winds [cold, dense Air rushing down slopes at night].

### The Wind Gradient and Wind Shear

When preparing for a Flight, analyzing the wind gradient is essential because friction from surface obstructions slows down wind speeds near the ground. This means wind velocity naturally increases with altitude.

The most critical wind hazard to anticipate is **wind shear**—a sudden, drastic change in wind speed or direction over a short distance. Wind shear can rapidly rob an Aircraft of Airspeed and lift, making it exceptionally dangerous during low-altitude maneuvers and final landing approaches.

## 3. Atmospheric Visibility Restrictions

Visibility establishes the baseline parameters for whether a Flight can legally and safely proceed under Visual Line of Sight [VLOS] regulations. Atmospheric visibility can deteriorate rapidly due to suspended water droplets or dry particles:

- **Mist vs. Fog:** Mist occurs when tiny water droplets limit horizontal visibility to between 1 and 5 kilometers. Fog develops when dense saturation drops horizontal visibility below 1 kilometer.
- **Radiation Fog:** Common on clear, calm winter nights in northern India, this forms as the ground loses heat through radiation, cooling the surrounding Air to its dew point.
- **Advection Fog:** This occurs when warm, moist Air moves horizontally over a cold surface, causing rapid condensation.
- **Haze and Smog:** In dryer or industrialized regions, blowing dust and industrial smog create a persistent layer of haze that scatters light, making it difficult to spot distant obstacles or incoming manned Aircraft.

## 4. The Impact of Rain, Moisture, and Storm Lifecycles

Rain and severe storms pose an immediate threat to the structural and electronic integrity of standard uncrewed Aircraft. Most consumer and commercial Drones possess low ingress protection [IP] ratings; prolonged exposure to rain can cause moisture to penetrate the Airframe, short-circuit electronic speed controllers [ESCs], and blind optical positioning sensors.

### The Lifecycle of a Thunderstorm

Thunderstorms are driven by intense convective activity. Understanding their three distinct phases helps Pilots identify when a system is too dangerous to approach:

1. **The Cumulus Stage:** Characterized by continuous warm updrafts that feed and build a growing cloud structure.

2. **The Mature Stage:** Marked by the onset of precipitation. Raindrops drag cold Air down, creating severe updrafts and downdrafts simultaneously. This stage generates intense wind shear, hazardous microbursts, heavy rain, and destructive lightning.
3. **The Dissipating Stage:** Occurs as cold downdrafts choke off the warm updrafts, causing the storm cell to lose energy and gradually decay.

Flying anywhere near active convective cells exposes a Drone to extreme updrafts that can violently alter its altitude, structural icing that ruins propeller aerodynamics, and catastrophic electrical failure from lightning strikes.

## 5. Decoding Weather Forecasting Data

Aviation forecasting transforms complex atmospheric theory into actionable operational data, enabling Pilots to make definitive go/no-go decisions well before arriving at a launch site. Commercial Drone Pilots should learn to check standardized alphanumeric Aviation weather reports:

- **METAR [Meteorological Aerodrome Report]:** A highly accurate snapshot of actual observed weather conditions at a specific Airport at a given time. It details current wind vectors, horizontal visibility, present weather phenomena, cloud layers, temperature, dew point, and local altimeter settings.
- **TAF [Terminal Aerodrome Forecast]:** A detailed statement of expected meteorological conditions at an Airport over a specified future period, usually spanning 12 to 30 Hours.
- **SIGMET [Significant Meteorological Information]:** High-priority Aviation advisories warning of severe, safety-critical atmospheric phenomena such as severe turbulence, heavy thunderstorms, or volcanic ash clouds.

Interpreting these data streams daily ensures that you never Fly blindly into deteriorating conditions, allowing you to align your Flight schedules safely within the operating envelope of your Aircraft.

# CHAPTER 21: RADIO TELEPHONY & COMMUNICATION

- **Communication Procedures**
- **Aviation Phraseology**
- **Emergency Communication**

# Chapter 21 Radio Telephony & Communication

Operating an Unmanned Aerial Vehicle [UAV] within shared civilian Airspace requires a clear line of communication. Whether you are Flying a large enterprise Drone near a controlled commercial Airport or navigating a mapping project in a busy industrial zone, you cannot afford to speak in casual, everyday language.

The **Radio Telephony [RT] & Communication** module details the specialized spoken codes, exact transmission habits, and emergency radio protocols required to coordinate smoothly with Air Traffic Control [ATC] and other Pilots.

## 1. Standard Aviation Communication Procedures

Radio communications in Aviation are handled on high-frequency or very high-frequency [VHF] radio bands. Because multiple Aircraft, ground teams, and controllers frequently share a single radio frequency, strict discipline is necessary to prevent transmissions from clipping or blocking each other.

**The Four Cardinal Rules of Radio Discipline**

- **Listen Before You Talk:** Before keying your microphone button, listen to the frequency for at least a few seconds. Interrupting an ongoing transmission between ATC and a commercial Airliner can cause dangerous confusion.
- **Think, Then Speak:** Know exactly what you are going to say before you transmit. Do not hold down the push-to-talk button while pausing or using filler words like "uh" or "um."
- **Be Clear and Concise:** Keep your messages as short as possible while conveying all necessary information. Use standard phrases so anyone on the frequency understands you instantly.

- **The Read-Back Requirement:** Whenever ATC issues a specific clearance, constraint, or coordinate shift, you must repeat the instructions back to the controller verbatim, ending with your specific Drone Pilot or station call sign. This confirms to the controller that you heard and understood the directive perfectly.

## 2. Standard Aviation Phraseology & Phonetics

To prevent misheard numbers or letters due to static, engine noise, or regional accents, the Aviation world relies on a standardized language developed by the International Civil Aviation Organization [ICAO].

**The NATO Phonetic Alphabet**

Every letter transmitted over an Aviation radio frequency must be spelled out using its designated phonetic word. For a Drone Pilot, this is most frequently used when stating your Aircraft's Unique Identification Number [UIN] or station ID.

A - Alpha

B - Bravo

C - Charlie

D - Delta

E - Echo

F - Foxtrot

G - Golf

H - Hotel

I - India

J - Juliet

K - Kilo

L - Lima

M - Mike

N - November

O - Oscar

P - Papa

Q - Quebec

R - Romeo

S - Sierra

T - Tango

U - Uniform

V - Victor

W - Whiskey

X - X-ray

Y - Yankee

Z - Zulu

If your Drone's UIN ends in **U-5-C**, you would pronounce it over the radio as **"Uniform, Fife, Charlie."** Note that the number five is spoken as *fife* and the number nine is spoken as *niner* to keep them from sounding like other words over a static-heavy transmission.

**Standard Operational Phraseology**

Casual words like "Yes," "No," and "Okay" are banned on Aviation frequencies. Instead, Pilots must use specific phraseology:

- **Roger:** I have received all of your last transmission. [Crucially, "Roger" does *not* mean "Yes" or "I comply"—it simply means "I received your message."]
- **Wilco:** I understand your message and **Will Co**mply with it.
- **Say Again:** Used to request that a controller or Pilot repeat their last transmission. [Never use the word "Repeat" on an Aviation radio, as that word is historically reserved for military artillery commands].

- **Affirmative / Negative:** Used instead of "Yes" or "No" to prevent confusion over static-filled links.
- **Acknowledge:** Let me know that you have received and understood this message.

## 3. Emergency Communication Protocols

When an inFlight crisis occurs—such as a critical battery failure, a complete loss of control link, or an incoming crewed helicopter Flying straight toward your operational grid—communication priorities change instantly.

Aviation protocols recognize two distinct tiers of radio emergency calls. Both must be broadcast on the active local traffic frequency, or on the international Aviation distress frequency [**121.5 MHz**].

**Distress Calls: MAYDAY**

The word **MAYDAY** [derived from the French *m'aider*, meaning "help me"] is reserved strictly for situations involving **immediate, catastrophic danger to an Aircraft or human life**, where immediate assistance is required. Examples include a structural Drone breakdown over a crowded area or an out-of-control Flyaway heading into a commercial Airport approach path.

A standard distress transmission follows a set structure:

1. **The Distress Signal:** Spoken clearly three times in a row: **"MAYDAY, MAYDAY, MAYDAY."**
2. **Identification:** Your Pilot call sign or Drone UIN.
3. **Position:** Your exact current location, coordinates, or altitude.
4. **Nature of the Emergency:** A brief description of what is failing [e.g., "Complete engine failure, unguided descent"].
5. **Intentions:** What you are doing to handle the risk [e.g., "Attempting ditching in uninhabited green zone"].

### Urgency Calls: PAN-PAN

The phrase **PAN-PAN** [from the French *panne*, meaning a mechanical breakdown] is used for **urgent situations concerning the safety of an Aircraft, vehicle, or person, but where there is no immediate danger to life**. Examples for a Drone team include a rapidly degrading battery cell that requires an unscheduled emergency landing, or spotting a brush fire from your payload camera that needs to be reported to authorities.

An urgency transmission follows a similar structure, beginning with the code words spoken three times: **"PAN-PAN, PAN-PAN, PAN-PAN,"** followed by your identification, current position, the nature of the urgent issue, and your intended plan of action to resolve it safely.

## 4. Summary Guide to Radio Transmission Priority

Radio frequencies have a strict legal hierarchy. You must yield the Airwaves instantly if a higher-priority transmission begins:

1. **Distress Communications [MAYDAY]:** Outranks all other traffic on the radio network. Complete radio silence must be maintained by all other stations.
2. **Urgency Communications [PAN-PAN]:** Second in priority. Takes precedence over all routine Flights and weather updates.
3. **Air Traffic Control Clearances:** Standard operational instructions, runway updates, and positioning vectors.
4. **Routine Communications:** Standard weather observations, Flight plan notifications, and regular ground coordinates.

**Radio Telephony & Communication**

- Communication Procedures
- Aviation Phraseology
- Emergency Communication

# CHAPTER 22: DRONE MAINTENANCE

- Preventive Maintenance
- Troubleshooting
- Battery Management
- Firmware Updates

# Chapter 22 Drone Maintenance

Proper maintenance is what separates a professional Drone operation from an amateur one. Because Unmanned Aerial Vehicles [UAVs] rely on high-speed moving parts, complex electronic circuitry, and chemical power sources, they experience continuous physical wear and vibrations during every Flight.

The Drone Maintenance module details how to establish pre- and post-Flight care routines, troubleshoot common field issues, manage Lithium-Polymer [LiPo] batteries safely, and execute firmware updates without bricking your equipment.

## 1. Preventive Maintenance: The Proactive Routine

Preventive maintenance is the practice of systematically inspecting and servicing your Drone to detect and resolve minor structural defects before they escalate into catastrophic mid-Air failures.

**Propeller Care and Inspection**

Propellers are your Drone's only source of lift and control; they require absolute structural integrity.

- **Micro-cracks and Nicking:** Before every launch, physically run your fingers down the edges of each propeller blade. Look for hAir-line fractures, chips, or bending caused by minor debris strikes. A damaged propeller experiences aerodynamic imbalance, which causes heavy vibrations that can destroy motor bearings and confuse the onboard Flight controller.
- **Cleaning Protocol:** Wipe down blades using a damp microfiber cloth to clear away dust, sap, and insect residue that alter the aerodynamic profile of the Airfoil.

- **Propeller Matching:** Never mix different brands or aging generations of propellers on the same Drone frame, as minor pitch variations will force the motors to work unevenly.

**Structural and Motor Integrity Checks**

- **The Chassis and Fasteners:** Inspect the structural arms, landing gear, and central body frame for deep stress cracks, especially around joints and motor mounts. Use a precision torque screwdriver to verify that all mounting screws are snug, as Drone vibrations naturally loosen fasteners over time.

- **Motor Bearing Checks:** With the battery disconnected, manually spin each brushless motor with your hand. The rotation should feel completely smooth, quiet, and free of resistance. If you hear a gritty, scratching noise or feel a grinding sensation, the motor bearings are contaminated with dirt or worn down and must be replaced immediately.

## 2. Troubleshooting Common Field Failures

Even with rigorous preventive care, components can experience unexpected glitches in the field. A Remote Pilot must know how to quickly isolate a failure using the Drone's telemetry system and onboard sensor readouts.

**Sensor Calibration Issues [IMU and Compass]**

The Inertial Measurement Unit [IMU] and the internal compass act as the Drone's inner ear. When they go out of alignment, the Drone loses its sense of balance and heading.

- **Symptom:** The Flight application displays an "IMU Error" or "Compass Interrupted" warning, or the Drone drifts aggressively to one side while hovering in a no-wind environment.

- **The Fix:** Move the Drone away from large metallic objects, concrete rebar, cars, or heavy power lines, which distort the Earth's natural magnetic field. Place the Drone on a perfectly level surface and execute a complete IMU and compass calibration sequence through your ground station software.

**Signal Loss and Radio Frequency Disruption**

- **Symptom:** The video downlink stream stutters, freezes, or drops entirely while the Drone is well within its specified legal operating range.
- **The Fix:** Verify your ground control station antenna orientation. Antenna signals travel outward perpendicular to the flat length of the antenna rod; pointing the sharp tips of your antennas directly at the Drone actually places the Aircraft in a reception dead zone. Adjust the paddles so the flat, wide faces point directly toward your Aircraft's operating position.

## 3. Battery Management: LiPo Safety Protocols

Most professional and enterprise Drones run on Lithium-Polymer [LiPo] battery packs. While LiPo batteries offer incredible power-to-weight ratios, they are volatile chemical cells that require strict adherence to handling and safety protocols.

**Charging and Discharging Windows**

- **The Golden Rule of LiPo Voltage:** A standard individual LiPo cell has a nominal voltage of 3.7V, a maximum fully-charged voltage of 4.2V, and a hard safety floor of 3.0V. Forcing a cell below 3.0V under load permanently damages the internal chemistry, causing rapid cell degradation or immediate battery failure.
- **Storage Voltage:** Never store your Drone batteries completely full [4.2V per cell] or completely empty [3.3V per cell] for more than a couple of days. If batteries are left fully charged, they will quickly begin to swell and lose capacity. Always discharge or charge them to a stable "Storage Level" of roughly **3.80V to 3.85V per cell** using a smart balance charger.

**Detecting and Managing Damaged Batteries**

- **Physical Swelling ["Puffing"]:** If a battery casing appears puffy, bloated, or swollen, the internal chemical layers have broken down, producing hazardous flammable gas. **Cease using a**

**swollen battery immediately.** Continued use exposes the pack to high internal heat, which can lead to a thermal runaway event—resulting in a violent chemical fire that cannot be put out by normal water extinguishers.

- **Safe Disposal:** To safely discard a compromised LiPo battery, discharge it completely down to 0.0V using a dedicated battery discharger or a light-bulb load bank, then drop it off at an authorized e-waste recycling center.

## 4. Firmware Updates: Keeping the Code Current

Firmware is the internal operating system running on your Drone's Flight computer, camera gimbals, Remote controller, and smart battery management units. Manufacturers regularly push firmware patches to update Airspace maps, patch software vulnerabilities, and add critical safety parameters.

### Pre-Update Safety Prerequisites

Executing a firmware update incorrectly can corrupt the device's internal flash memory, rendering the Drone inoperable [commonly known as "bricking"]. Always adhere to these safety rules before clicking update:

- **Battery Check:** Ensure that both the Drone battery and your ground control station controller are charged to **at least 50% capacity** [ideally 100%]. If either device loses power midway through a core firmware flash, it can corrupt the system completely.
- **Remove the Propellers:** This is the most vital safety step in Drone maintenance. During firmware updates, the Flight controller resets its internal input/output pins. This can occasionally trigger a bug that arms the electronic speed controllers, spinning the motors up to full throttle unexpectedly. **Always take the propellers off before updating.**
- **Post-Update Recalibration:** Major firmware updates frequently erase old sensor offsets. After completing a system update, you

must perform a fresh calibration of your IMU, compass, and Remote control sticks before launching your next mission.

- Preventive Maintenance
- Troubleshooting
- Battery Management
- Firmware Updates

# CHAPTER 23: SIMULATOR TRAINING

- **Simulator Setup**
- **Virtual Missions**
- **Emergency Practice**

# Chapter 23 Simulator Training

Before you ever place a real Drone onto a live launch pad or turn its propellers in shared Airspace, you must develop intuitive muscle memory on the ground. Professional Aviation training has long relied on simulated environments to build competence, and unmanned Aviation follows the exact same philosophy.

The Simulator Training module bridges the gap between theoretical classroom learning and physical field deployments. By practicing in a zero-risk virtual space, you can master control inputs, complete complex Flight paths, and experience catastrophic emergencies without risking expensive equipment or public safety.

## 1. Simulator Setup and Configuration

A quality simulation experience depends heavily on how accurately your physical controls mirror a real Flight environment. Training with a standard keyboard, mouse, or video game controller is counterproductive, as it builds incorrect muscle memory that can lead to field accidents.

### Hardware Interfacing

The gold standard for simulator setup is connecting your actual Drone **Ground Control Station [GCS]** or RC transmitter directly to a computer using a USB interface cable or a wireless trainer dongle. This setup ensures that the stick tension, physical travel distance, and switch layouts match your operational equipment exactly.

### Control Mode Selection

Before spinning up your virtual motors, you must verify your stick configuration menu. The vast majority of commercial Drone Pilots in

India operate under **Mode 2**, which positions the primary controls as follows:

- **Left Stick:** Controls Throttle [vertical climb and descent] and Yaw [nose rotation left or right].
- **Right Stick:** Controls Pitch [forward and backward horizontal slide] and Roll [sideways banking left or right].

**Software Calibration**

Once connected, you must complete a full axis calibration within the simulation software. This step sets the absolute center points and maximum movement boundaries for your control sticks, preventing issues like virtual joystick drift or erratic control inputs during training modules.

## 2. Virtual Missions: Building Spatial Awareness

Virtual missions are structured scenarios designed to develop a Pilot's depth perception, orientation tracking, and spatial awareness across various environments and weather patterns.

**Core Maneuvering Drills**

- **The Static Hover Challenge:** Keeping the virtual Drone locked inside a tight, floating target box at a specific altitude. This drill helps you practice micro-adjustments on the sticks to counteract simulated wind drifts.
- **The Nose-In Orientation Drill:** Hovering the Drone with its camera nose pointing directly back at you. When a Drone faces the Pilot, horizontal control inputs are completely reversed—pushing the roll stick right causes the Drone to slide to your left. Simulators allow you to master this mental inversion safely.
- **The Figure-Eight Path:** Flying a continuous, fluid infinity pattern around two virtual pylons. This advanced drill requires coordinating throttle, pitch, roll, and yaw inputs simultaneously to keep the Aircraft moving smoothly.

**Industrial Environment Simulations**

Advanced training platforms let you transition from open fields to complex industrial environments. You can practice Flying tight grid lines over a virtual construction site, navigating around high-voltage power lines, or managing tight clearances between tight structural columns on an automated inspection mission.

## 3. Emergency Practice: Training for the Worst

The greatest value of simulator training is the ability to practice dangerous, life-threatening Flight failures that are impossible to test safely with a real Aircraft. Simulators allow instructors to instantly trigger random system failures at the click of a button.

**Total GPS Failure and ATTI Mode Overrides**

- **The Scenario:** The simulator suddenly cuts the Drone's satellite connection midway through a high-altitude mission, replicating heavy magnetic or ionospheric interference.
- **The Emergency Behavior:** The Drone loses its ability to anchor itself automatically in space. If a wind gust hits the Aircraft, it will drift rapidly and continuously in the direction of the wind.
- **The Recovery Protocol:** The Pilot must immediately switch the Flight mode to **ATTI [Attitude] Mode** and manually counter the wind drift using the right stick. You must visually track the Aircraft and steer it back to a safe manual landing without relying on automated holding systems.

**In-Flight System Criticalities**

- **Control Link Loss [Flyaway Prevention]:** Practicing what to do if your screen blank-outs or your Remote disconnects. You will learn to trust and monitor your automated **Return-to-Home [RTH]** safety triggers, ensuring your pre-programmed RTH clearing altitude is set higher than local obstacles.
- **Sudden Battery Voltage Drops:** Simulating a rapid battery cell failure that forces the Drone into an immediate, unpreventable

land-in-place sequence. Pilots practice steering the descending Drone away from virtual crowds or water hazards into a clear, flat clearance zone during its final seconds of powered Flight.

## 4. Simulator Progress Metrics

To maximize the benefits of your ground school simulation time, monitor these operational metrics before clearing yourself for a live field deployment:

- **Crash-Free Flights:** You should be able to complete at least 20 consecutive missions across various weather variations without a single structural collision or hard landing.
- **Orientation Recovery:** You should be able to instantly recover control of the Drone from a distance when its nose orientation is pointing away from your line of sight during blind wind gusts.
- **Muscle Memory Continuity:** Your stick responses during emergency cut-outs should feel smooth and deliberate, replacing panicked, jerky inputs with measured, corrective adjustments.

# CHAPTER 24: PRACTICAL FLIGHT TRAINING

- **Hovering**
- **Takeoff**
- **Landing**
- **Navigation**
- **Mission Flying**

# Chapter 24 Practical Flight Training

The final transition from a simulator to a certified field Pilot culminates on the Flight line. While simulators construct the foundational muscle memory, they cannot completely replicate the real-world variables of live field environments—such as shifting ground-effect Air currents, unpredictable wind shears, or the physiological pressure of Piloting expensive hardware.

The Practical Flight Training module translates your ground school theory into real-world Flight mastery, moving step-by-step from base take-off routines to professional mission execution.

## 1. Ground Preparation and Takeoff Procedures

A successful Flight line deployment begins long before the motors spin up. Ground preparation establishes a safe, controlled workspace to minimize structural risks during the initial transition into the Air.

### Site Assessment and Launch Pad Setup

The Remote Pilot must establish a clear take-off and landing zone with a minimum radius of 3 meters, free from loose debris, tall grass, and overhead obstructions like branches or utility lines. Using a heavy, weighted physical landing pad is highly recommended; it prevents the high-speed downdrafts of the propellers from kicking up fine dust, sand, or loose dirt that can clog motor bearings and scratch sensitive camera optics.

### Executing the Takeoff

Once the comprehensive pre-Flight checklist is completed and a solid GPS lock is secured, the Pilot executes the takeoff sequence.

- **The Manual Arming Command:** The Pilot pulls both control sticks down and inward [or outward, depending on the manufacturer's layout] to arm the brushless motors at an idle speed.

- **The Initial Ascent:** The Pilot applies a smooth, steady upward pressure on the throttle stick, accelerating the Drone quickly through the "ground effect" zone—a layer of turbulent Air trapped between the Drone and the ground within the first 3 feet of altitude. The Aircraft should climb cleanly to an eye-level holding altitude of roughly 6 to 10 feet AGL.

## 2. Mastering the Static Hover

Hovering is the absolute baseline of all multi-rotor operations. It represents a state of aerodynamic equilibrium where the Aircraft remains locked at a precise altitude and geographical coordinate.

### Wind Correction and Orientation Awareness

Even with a stable GPS lock, micro-currents of wind will cause minor movements. The Pilot must learn to make subtle, deliberate stick corrections rather than jerky, over-reactive movements that cause the Drone to oscillate erratically.

The core of hover training involves mastering **spatial orientation cycles**:

- **Tail-In Hovering:** Operating with the rear of the Drone facing you. This is the easiest orientation because the Drone's movements map perfectly to your control sticks.
- **Side-In Hovering:** Rotating the Drone 90 degrees so its side profile faces you. This tests your depth perception as you manage lateral tracking.
- **Nose-In Hovering:** Rotating the Drone 180 degrees so it faces you directly. In this position, your horizontal control inputs are inverted, requiring focused mental coordination to slide left when the stick is pushed right.

## 3. Basic and Advanced Navigation Flight Paths

Once you can maintain a stable hover in any orientation, you progress to executing structured, geometric Flight profiles to build complete control over the Aircraft's three axes of motion.

### Core Geometric Maneuvering Drills

- **The Square Circuit:** The Pilot flies the Drone along a strict square trajectory, stopping at each corner to pivot the nose 90 degrees using the yaw stick before moving forward along the next edge. This drill reinforces precise start-and-stop control.

- **The Continuous Circle:** Flying a perfect circle around a fixed ground coordinate while keeping the camera nose pointed directly toward the center. This requires coordinating pitch, roll, and yaw inputs simultaneously.

- **The Horizontal Figure-Eight:** An advanced drill where the Drone traces an infinity path across the sky. This exercise trains the Pilot to manage continuous transitions in orientation and momentum smoothly.

## 4. Mission Flying and Enterprise Profiles

Mission Flying moves beyond simple manual steering to replicate the exact Technical Flight patterns required across various commercial industries.

### Industrial Flight Patches

- **The Grid Matrix [Aerial Mapping]:** The Pilot configures the Drone to Fly a precise, automated "lawnmower" grid pattern over a specified area. The Aircraft must maintain a constant altitude and speed while triggering the camera sensor at mathematically calculated intervals to ensure adequate image overlap for 3D modeling.

- **The Corridor Track [Linear Inspections]:** Used extensively for monitoring cross-country pipelines, railway beds, and high-voltage electrical grids. The Pilot commands the Drone to follow a narrow, linear path, keeping the payload camera focused sharply on the asset while staying clear of nearby structures.

- **The Point-of-Interest Orbit [Structural Inspections]:** Executing an automated or manual high-precision circle around a vertical structure, like a telecom tower or wind turbine. The camera must remain locked onto the asset faces to capture high-definition structural imagery from all angles.

## 5. Controlled Landing Protocols

A Flight is not truly successful until the Aircraft is safely powered down on the landing pad. The landing phase demands high focus, as the Drone must re-enter turbulent lower Air currents close to ground hazards.

**The Landing Sequence**

- **The Approach:** The Pilot brings the Drone to a stable hover directly over the center of the landing pad at an altitude of 5 feet, verifying that the immediate landing zone remains clear of personnel or animals.
- **The Descent:** The Pilot applies a gentle, continuous downward pressure on the throttle stick. The descent must be slow and steady; pulling the throttle down too aggressively can cause the Drone to enter a dangerous aerodynamic state known as "vortex ring state," where the Aircraft sinks into its own turbulent downwash and loses control.
- **Disarming:** Once the landing gear legs make firm contact with the ground pad, the Pilot holds the throttle stick down completely for three seconds or executes the manual disarm command. This action cuts power to the electronic speed controllers, stopping the propellers instantly before any post-Flight data retrieval begins.

Practical Flight Training

- Hovering
- Takeoff
- Landing
- Navigation
- Mission Flying

# Chapter 25: Risk Assessment & Safety Management

- **Hazard Identification**
- **Human Factors**
- **Safety Systems**
- **Emergency Procedures**

# Chapter 25 Risk Assessment & Safety Management

Every professional Drone deployment carries a level of inherent risk. Unlike traditional Aircraft with built-in redundancies and onboard Flight crews, an Unmanned Aerial Vehicle [UAV] relies entirely on its ground-based Pilot to navigate unexpected structural errors, human limitations, and unpredictable field environments.

The **Risk Assessment & Safety Management** module establishes the ultimate operational framework for professional Flyers. It focuses on identifying hidden hazards, managing psychological errors, leveraging Technical safety systems, and deploying rapid emergency protocols to minimize liabilities and safeguard public property.

## 1. Hazard Identification and Risk Mitigation

A hazard is any real or potential condition that can cause injury, illness, or death to personnel, or damage to equipment and property. Risk is the calculation of that hazard's severity combined with its probability of occurrence.

**The Pre-Flight Site Assessment**

Before unpacking any Flight gear, a Remote Pilot must perform a rigorous, physical assessment of the surrounding operational site to look for specific hazard categories:

- **Physical Hazards:** Overhead high-voltage power lines, communication towers, construction cranes, tall trees, and local structural facades.
- **Operational Hazards:** Dense gatherings of people, moving vehicle traffic corridors, and active livestock.

- **Environmental Hazards:** Approaching wind fronts, localized thermal updrafts, or heavy bird nesting areas nearby.

**The Risk Matrix Framework**

To structure safety decisions, professional operators apply a standard Risk Matrix. This tool helps you evaluate whether a Flight should proceed or be aborted based on two axes:

- **Probability:** How likely is an accident to happen? [Rated from Frequent to Improbable].
- **Severity:** How bad will the damage be if it does happen? [Rated from Catastrophic to Negligible].

If a hazard falls into a high-risk zone—such as mapping a construction site directly next to a school during recess—the Pilot must implement mitigation strategies [e.g., rescheduling the Flight to early morning Hours when the school is empty] to bring the total risk level down to an acceptable baseline before launching.

## 2. Managing Human Factors in Aviation

Human factors refers to the study of how people interact with their equipment, environments, and operational data. Over 80% of documented Aviation accidents are traced back to human error rather than mechanical breakdown.

**The IMSAFE Personal Assessment Checklist**

A certified Remote Pilot must objectively evaluate their own physiological and psychological health before taking control of an Aircraft. The Aviation industry relies on the **IMSAFE** checklist to catch hidden performance vulnerabilities:

- **I - Illness:** Am I suffering from any current illness, cold, or physical ailment that impAirs my reflexes?
- **M - Medication:** Have I taken prescription or over-the-counter drugs that cause drowsiness or slow down my processing time?

- **S - Stress:** Am I distracted by heavy professional, financial, or personal psychological stressors?
- **A - Alcohol:** Have I consumed alcohol within the last 8 to 24 Hours? [Aviation rules maintain a zero-tolerance policy for intoxication].
- **F - Fatigue:** Am I suffering from a lack of sleep or physical exhaustion that ruins my tracking abilities?
- **E - Emotion:** Am I feeling unusually anxious, angry, or rushed to finish the project?

### Overcoming Hazardous Attitudes

Pilots must also monitor their mental state for dangerous behavioral patterns that lead to bad decision-making in the field:

- **Macho Attitude ["I can handle anything"]:** Taking unnecessary risks to impress onlookers. *The antidote: Taking chances is foolish.*
- **Impulsivity ["Do something quickly"]:** Reacting instantly to an error without thinking through the safety checklist. *The antidote: Not so fast, think first.*
- **Resignation ["What's the use?"]:** Feeling helpless when an emergency occurs and giving up control. *The antidote: I am not helpless; I can make a difference.*

## 3. Technical Safety Systems & Built-In Redundancies

Modern enterprise Drones feature layers of automated software and hardware defenses designed to step in when external conditions degrade or control links break down.

### Geofencing Technology

Geofencing uses live GPS tracking combined with internal databases to build virtual, invisible walls around the Aircraft.

- **Horizontal Geofencing:** Prevents the Drone from drifting laterally into restricted spaces, such as active Airport perimeters or highly sensitive industrial zones.
- **Vertical Geofencing:** Acts as a hard software ceiling that stops the Drone from exceeding the legal 400-foot AGL limit, keeping it clear of traditional manned Aviation lanes.

**Fail-Safe Automation**

- **Return-to-Home [RTH]:** If the radio control link between your ground station and the Drone breaks completely due to terrain interference or battery drop-outs, the Aircraft pauses for a few seconds. It then climbs to a pre-set safe clearing altitude, navigates back to its original takeoff coordinates via GPS, and executes an automated landing.
- **Obstacle Avoidance Systems:** Utilizing combinations of stereo vision sensors, infrared time-of-Flight arrays, and LiDAR sensors, modern Drones scan their immediate Flight path in 360 degrees. If an obstacle like a tree branch or building corner is detected, the Drone will automatically brake, hover in place, or intelligently chart a path around the hazard.

## 4. Emergency Procedures: Action Protocols

When an inFlight crisis hits, there is no time to read through a manual. A Pilot must react with fast, structured emergency protocols to minimize injury and property damage.

**Handling an In-Flight Battery Fire [Thermal Runaway]**

If your telemetry screen flashes a critical battery temperature warning or you see a trail of thick white or grey smoke venting from your Drone, you are experiencing a structural chemical breakdown.

- **Action:** Immediately steer the Drone down and away from buildings, vehicles, and crowds. Execute a hard landing in an open green space, dirt patch, or gravel zone. Keep bystanders away from the landing site, as a burning lithium pack releases highly toxic gas and can explode violently.

## Managing a Loss-of-Control Link [Flyaway Event]

If the Drone stops responding to your control sticks and begins drifting away randomly without triggering its automatic RTH protocol, it is considered a "Flyaway."

- **Action:** Toggle your Flight mode switch rapidly from GPS mode to **ATTI [Manual] Mode** to try and break the software loop and regain manual stick control. If the Drone continues to drift toward controlled Airspace [like an Airport corridor], you must immediately notify local Air Traffic Control [ATC] and law enforcement, providing them with your last known location, current altitude, heading, and estimated remaining battery life.

## Managing Airspace Incursions by Crewed Aircraft

If you hear the low rumble of an approaching low-Flying medical helicopter or crop-duster plane, you must treat it as an immediate safety priority.

- **Action:** Crewed Aircraft *always* maintain the absolute right of way in shared Airspace. You must immediately drop your Drone's altitude below 100 feet AGL or land completely, clearing the sky entirely until the crewed Aircraft leaves the operational area.

Risk Assessment & Safety Management

- Hazard Identification
- Human Factors
- Safety Systems
- Emergency Procedures

# PART VI

# AVIATION FUNDAMENTALS FOR DRONE PILOTS

# Chapter 26: Introduction to Aviation

- Aviation Industry Overview
- Air Traffic System
- Airports

# Chapter 26 Introduction to Aviation

Operating a commercial Drone means stepping out of the amateur Tech space and entering the professional world of civil Aviation. The moment your uncrewed Aircraft leaves the ground, it shares an integrated ecosystem with commercial Airliners, cargo jets, and military transport vehicles.

To safely exist within this space, a Remote Pilot must understand the broader Aviation industry, the structural layout of national Air traffic networks, and the physical configuration of Airport environments.

## 1. Aviation Industry Overview

The global Aviation industry is a highly organized transportation network divided into three major operational sectors. Drone Technology intersects directly with these existing pillars.

### Commercial Aviation

This sector includes scheduled passenger Airlines and international Air cargo networks that move millions of people and tons of freight daily. These Aircraft operate at high speeds and high altitudes, following strict, pre-planned Flight tracks. Because commercial Airliners require vast sections of protected Airspace during their take-off and final landing approaches, Drone Pilots must avoid these key corridors.

### Military Aviation

This branch includes the Aerial operations conducted by national defense forces, including fighter jets, heavy transport Aircraft, and surveillance platforms. Military Flights often execute high-speed, low-altitude training maneuvers. Drone Pilots working near defense properties or temporary military exercise zones must stay alert to handle sudden, fast-moving traffic.

**General Aviation [GA]**

General Aviation represents the diverse backbone of civil Flight. It encompasses all non-scheduled, non-military civilian operations, including private charter Flights, medical evacuation helicopters, Flight school training planes, and agricultural crop-dusters.

Because General Aviation Aircraft frequently operate at lower altitudes—often dipping below 1,000 feet Above Ground Level [AGL] for surveying, photography, or landing practice—**this sector interacts most frequently with Drone operations.** Drone Pilots must understand that small helicopters and low-Flying planes can appear suddenly, and the Drone must always yield the right of way.

## 2. The Air Traffic System and Airspace Classes

To maintain order and prevent mid-Air collisions, the sky is managed by a centralized network called the **Air Traffic System [ATS]**, which is overseen regionally by Air Traffic Control [ATC].

**Controlled vs. Uncontrolled Airspace**

Global civil Aviation authorities slice the atmosphere into distinct, invisible blocks of Airspace. These are broadly categorized into two operational types:

- **Controlled Airspace:** Airspace zones where structural Air traffic control services are actively provided. In these blocks, crewed Aircraft must maintain continuous two-way radio contact with controllers and secure explicit clearances before changing their speed, heading, or altitude. Drone operations inside controlled zones [such as Yellow or Red zones near cities and transit hubs] require formal ATC approval via the Digital Sky Platform.
- **Uncontrolled Airspace:** Airspace blocks where the volume of commercial traffic is low enough that active ATC steering is not mandatory. In this space, Pilots Fly using visual separation rules

["see and avoid"]. In India, the lower boundary of uncontrolled Airspace is designated as the **Green Zone**, extending up to **400 feet [120 meters] AGL**. Registered Drone Pilots can operate freely within this window without needing real-time radio clearances, provided they maintain visual line of sight.

## 3. Airport Layouts and Operational Environments

Airports are high-velocity, high-risk Aviation hubs where Aircraft transition between the ground and the sky. A Remote Pilot executing a project near an Airport perimeter must understand basic Airport architecture to identify potential hazards.

**The Runway Environment**

Runways are flat, cleared strips of asphalt or concrete reserved exclusively for Aircraft takeoffs and landings. They are structurally numbered based on their magnetic compass heading, rounded to the nearest ten degrees. For example, a runway pointing directly east [90 degrees] is labeled **Runway 09**, while a runway pointing west [270 degrees] is labeled **Runway 27**.

[Image diagram of standard Airport runway layout showing approach path and taxiways]

Drone Pilots must understand that the hazard zone does not end at the physical Airport fence. Aircraft on final approach are often Flying dangerously low to the ground for several kilometers aligned with the runway centerline.

**Taxiways and Aprons**

- **Taxiways:** Designed paths used by Aircraft to roll slowly between maintenance hangars, passenger terminals, and active runways. They are typically designated by letters [e.g., Taxiway Alpha, Taxiway Bravo].
- **The Apron [or Tarmac]:** The designated area where Aircraft park to load passengers, refuel, and undergo pre-Flight inspections.

While ground movements on aprons are slow, they generate high levels of radio interference, which can disrupt nearby Drone control frequencies.

**Navigational Hazards: The Funnel of Protection**

Airports are protected by a strict three-dimensional boundary known as the **Aerodrome Traffic Zone [ATZ]**. Under Indian Drone Rules, the area within a **5-kilometer radius** of any operational Airport runway is classified as a hard **Red Zone**.

This 5 km buffer acts as a vital safety funnel, ensuring that low-Flying commercial jets and descending turboprops have a completely clear Airspace volume, entirely free from uncrewed Aircraft interference.

# Chapter 26: Introduction to Aviation

- **Aviation Industry Overview**
- **Air Traffic System**
- **Airports**

# Chapter 26A Introduction to Aviation

Aviation is one of the most dynamic, highly regulated, and technologically advanced industries in the world. It serves as a vital economic engine, connecting global markets, driving international trade, and supporting millions of jobs. Moving past the initial challenges of the early 2020s, the global and Indian Aviation sectors have entered an intensive phase of modernization, marked by record-breaking Aircraft orders, infrastructure expansion, and a shift toward digital optimization.

To understand how traditional manned Aviation functions—and how unmanned systems (drones) integrate safely into the same skies—you must understand the three core elements: the **Aviation Industry**, the **Air Traffic System**, and the underlying **Airport Infrastructure**.

## 1. Aviation Industry Overview

The civil Aviation ecosystem is broadly divided into two operational segments: **Commercial Air Transport** (scheduled passenger and cargo Flights) and **General Aviation** (private Flying, training, and corporate charters).

India is currently the world's third-largest domestic Aviation market, experiencing massive structural evolution. The industry landscape is defined by distinct carrier models and robust growth vectors:

- **Low-Cost Carriers (LCCs):** LCCs dominate the domestic market, holding roughly 69% of total capacity. Spearheaded by carriers like IndiGo (which commands over 50% market share) and rapid entrants like Akasa Air, the LCC model focuses on high-frequency, point-to-point routes and single-class configurations to keep Air travel accessible.
- **Full-Service Carriers (FSCs):** FSCs offer premium tiered services, multi-cabin layouts, and long-haul international networks. The

consolidation of Air India under the Tata Group transformation program (Vihaan.AI) represents one of the largest corporate turnarounds in Aviation history, signaling a dual-hub long-haul strategy out of major metros.

- **Fleet Projections:** Reflecting long-term market confidence, Indian carriers have placed historic, multi-hundred Aircraft orders with manufacturers like Airbus and Boeing. The country is projected to require approximately 22,000 commercial Pilots by 2034 to sustain this capacity rollout.

## 2. The Air Traffic System: Managing the Skies

**Air Traffic Control (ATC) Divisions**

As an Aircraft progresses through its journey, responsibility shifts across three specialized ATC units:

1. **Tower Control (Local/Aerodrome Control):** Manages the physical runways, taxiways, and the immediate Airspace surrounding the Airport (typically a 5-mile/8-km radius). Tower controllers issue takeoff and landing clearances.
2. **Approach Control (Terminal Radar Approach):** Coordinates arriving and departing Aircraft within a busier, wider radius around major Airport hubs. They sequence Aircraft into neat arrival lines before handing them off to Tower Control.
3. **Area Control Center (ACC / En-Route Control):** Manages the high-altitude commercial Air corridors between cities and oceans. Airspace is divided into sectors; as a plane flies across the country, controllers pass communication seamlessly from one sector to the next.

## 3. Airports: The Ground Infrastructure

Airports are highly secure, logistically dense facilities divided strictly into two operational zones: Landside and Airside.

- **Landside:** Areas accessible to the general public, including check-in counters, parking blocks, and terminal lobbies. Innovations like

**DigiYatra** use decentralized ID networks and facial recognition at the gate entry to significantly reduce passenger processing bottlenecks.

- **Airside:** Highly restricted zones beyond passport and security checkpoints. This includes the **Apron** (where Aircraft park at gates for boarding and fueling), **Taxiways** (the transit lanes for Aircraft rolling to and from the runway), and the **Runways** themselves.

**Greenfield Expansion and Regional Hubs**

To relieve chronic congestion at major single-runway or dual-runway hubs like Mumbai's CSMIA or Delhi's IGI Terminal 1, the Aviation sector has rapidly scaled infrastructure under central frameworks like *PM Gati Shakti.*

The network of operational Airports in India has expanded dramatically over the past decade, climbing past 160 active sites. This growth has been driven by the **UDAN** (Regional Connectivity Scheme), which funds the reactivation of unserved regional Airstrips, and massive greenfield developments like the **Navi Mumbai International Airport** and the **Noida International Airport (Jewar)**. These facilities act as relief valves, introducing needed slots, dedicated Air-cargo logistics terminals, and upgraded ground infrastructure to handle the next generation of Air traffic.

# CHAPTER 27: AIR TRAFFIC CONTROL BASICS

- **ATC Functions**
- **Communication Flow**
- **Flight Coordination**

# Chapter 27 Air Traffic Control Basics

Air Traffic Control [ATC] functions as the central nervous system of civil Aviation. Its primary objective is to maintain a safe, orderly, and expeditious flow of Air traffic across the skies. As a professional Remote Pilot, you must understand how ATC coordinates movements, manages the flow of communication, and processes Flight data to ensure your Drone operations never disrupt crewed Aircraft.

## 1. Primary Functions and Structure of ATC

The overarching purpose of ATC is to prevent collisions between Aircraft in Flight, prevent collisions on the ground between Aircraft and structural hazards, and organize the movement of all Air transport. To manage this safely, ATC responsibilities are split into three distinct operational domains:

- **Tower Control [Local Control]:** Operating from the glass-walled cab at the top of the Airport tower, these controllers manage all traffic on the active runways and within the immediate vicinity of the Airport [usually up to a 5-mile radius]. They issue structural clearances for takeoffs and landings.
- **Approach Control:** This department manages arriving and departing Aircraft within a wider terminal area surrounding an Airport [typically up to a 30- to 50-mile radius]. Approach controllers use radar to sequence fast-moving Airliners into a tight, orderly line before handing them over to the Tower.
- **Area Control Center [ACC / En-Route]:** These centers handle the high-altitude en-route traffic Flying between major cities and transit hubs. They ensure Aircraft stay safely separated as they cruise across domestic and international Air corridors.

## 2. The Standardized Flow of Communication

To guarantee that critical safety instructions are never misheard, ATC uses a rigid, standardized communication flow. Radio frequencies are a shared resource; multiple Pilots are listening at any given time, meaning clarity and brevity are legally mandated.

**The Anatomy of an ATC Transmission**

Every single exchange between a Pilot [crewed or uncrewed] and an ATC controller must follow a structured four-part format to prevent confusion:

1. **The Call Sign of the Station Being Addressed:** You always state who you are talking to first [e.g., *"Mumbai Tower..."*].
2. **Your Specific Call Sign:** You identify yourself next using your Aircraft registration or station ID [e.g., *"...Drone Bravo-Delta-Four"*].
3. **The Core Message:** The specific request or notification [e.g., *"... requesting permission to launch from site alpha for a localized mapping profile up to 200 feet AGL"*].
4. **The Read-Back and Sign-Off:** If ATC issues an instruction, you must repeat the exact parameters back to them, followed by your call sign, to confirm you understood the directive.

**The Separation of Instructions:** If an ATC controller tells you, "Maintain hover at one hundred feet, traffic is a low-Flying helicopter crossing from East to West," you do not just say "Okay." You must respond: *"Maintain hover at one hundred feet, copy traffic, Drone Bravo-Delta-Four."*

## 3. Flight Coordination and Digital System Integration

Traditional Aviation relies on paper Flight strips and verbal phone handoffs between controllers to track Aircraft. For Drone Pilots, Flight coordination has been digitized and automated to handle the high volume of unmanned missions seamlessly.

## Digital Integration and the NPNT Protocol

In India, the coordination between a Remote Pilot and ATC is managed primarily through the **Digital Sky Platform** using a system called **No Permission, No Takeoff** [NPNT]. Instead of calling a control tower on a phone to ask for clearance to Fly in a Green or Yellow zone, the process is streamlined through software integration:

- **Submission:** The Remote Pilot uploads a digital Flight plan specifying the exact geographic coordinates, date, time window, and maximum altitude of the planned mission.
- **Automated Cross-Referencing:** The platform automatically compares the request against current live Airspace restrictions, temporary Flight restrictions [TFRs], and Airport proximity limits.
- **The Clearance Artifact:** If approved, the system generates an encrypted digital token [the permission artifact]. This file is loaded into the Drone's firmware, unlocking the motors for that specific time and location.

## Manual Coordination Rules for Exceptional Circumstances

If you are authorized to operate a Drone inside a controlled **Yellow Zone** or near a localized industrial corridor that interfaces directly with an active approach path, digital clearances are frequently backed up by direct manual coordination:

- **Pre-Flight Notification:** The Pilot must establish contact with the concerned ATC supervisor at least 24 Hours prior to the Flight to review the mission scope.
- **The Land-Immediately Directive:** If an unexpected Aerial emergency arises [such as an Air-ambulance helicopter executing an unscheduled off-field landing near your site], ATC will broadcast an immediate recall instruction on the designated frequency. The Remote Pilot is required to land the Drone immediately, terminate the mission, and report compliance to the controller.

# CHAPTER 28: AVIATION SAFETY CULTURE

- Safety Principles
- Incident Prevention
- Human Performance

# Chapter 28 Aviation Safety Culture

Aviation is not inherently safe. It is, by its very nature, unforgiving of carelessness, ignorance, or neglect. Over a century of crewed Flight has proven that safety cannot be achieved through reactive rules alone; it requires a proactive, deeply ingrained **Aviation Safety Culture**.

For Remote Pilots, adopting this culture means moving past a casual Tech-hobbyist mindset and embracing the systematic safety disciplines used by commercial Airlines and military Flight crews worldwide.

## 1. Core Safety Principles: Just Culture vs. Blame Culture

A professional Aviation environment thrives on a baseline principle: human beings will occasionally make mistakes. How an organization handles those mistakes determines its true safety baseline.

### Just Culture

The global standard for Aviation safety is a **Just Culture**. In a Just Culture, Pilots and ground crews are actively encouraged to report their own errors, near-misses, or equipment glitches without fear of immediate punishment or retaliation. The primary goal is to learn *why* the mistake happened so the system can be updated to prevent a future accident.

However, a Just Culture draws a clear, uncompromising line between honest human errors and willful violations:

- **Honest Error [Protected]:** A Pilot miscalculates a wind gust gradient during a difficult landing, causing a minor hard landing, and immediately logs the incident. This is treated as a learning opportunity.

- **Willful Violation [Punished]:** A Pilot intentionally flies a Drone into a restricted Airport approach path to get scenic video footage, knowingly violating the law. This is reckless misconduct and results in strict legal prosecution.

**Blame Culture**

In contrast, a **Blame Culture** punishes any employee who makes a mistake. This environment forces teams to hide their errors, cover up near-misses, and falsify Flight logs to avoid losing their jobs or credentials. This silence guarantees that hidden mechanical faults or bad operational processes go uncorrected until they cause a catastrophic, public failure.

## 2. Systematic Incident Prevention Techniques

Incident prevention relies on catching and breaking the "chain of events" before it culminates in an accident. Disasters are rarely caused by a single isolated failure; they are almost always the result of multiple minor errors stacking up sequentially.

**The Swiss Cheese Model of Accident Causation**

Developed by cognitive psychologist James Reason, the Swiss Cheese Model visualizes an organization's safety defenses as a series of sequential slices of cheese placed side-by-side.

[ Hazard ] ---> [ ] ---> [ ] ---> [ ] ---> [ ] ---> [ ACCIDENT ]

Slice 1 Slice 2 Slice 3 Slice 4

Training Checklist Maintenance Pilot State

Each slice represents a layer of protection designed to stop a hazard from becoming an accident [such as rigorous training, mandatory pre-Flight checklists, strict hardware maintenance, and physical Pilot health assessments]. Normally, the holes in the slices do not line up.

An accident occurs only when a specific set of circumstances allows holes in *every single slice* to align perfectly, letting the hazard pass straight through all defenses:

1. **Hole 1 [Maintenance]:** The team skips a routine motor bearing check to save time.
2. **Hole 2 [Environment]:** Sudden, unforecasted turbulent wind gusts hit the project site.
3. **Hole 3 [ Pilot Fatigue]:** The Remote Pilot operates on three Hours of sleep, slowing their reflexes.
4. **The Result:** The worn motor bearing seizes under the stress of the wind, and the fatigued Pilot fails to transition to manual ATTI mode quickly enough, causing a high-speed crash into a crowd.

By enforcing strict, independent checks at every stage, you plug the holes in the cheese, ensuring that even if one layer fails, the next layer catches the error.

## 3. Optimizing Human Performance on the Flight Line

Human performance refers to how human capabilities and limitations impact the safety and efficiency of Flight operations. Remote Pilots face unique psychological challenges because they sit outside the physical cockpit, losing the immediate sensory feedback of motion, balance, and spatial sound.

**The Threat and Error Management [TEM] Framework**

Professional Drone crews use the Threat and Error Management framework to analyze risks in real time during field operations. The system splits field variables into two categories:

- **Threats:** External situations that occur outside the direct control of the Flight crew, which increase operational complexity. Examples include a sudden flock of birds entering the area, dropping temperatures draining batteries faster than expected, or a curious bystander walking up to talk to the Pilot during a delicate maneuver.

- **Errors:** Internal mistakes or miscalculations made directly by the Flight crew that reduce safety margins. Examples include entering the wrong Return-to-Home altitude into the ground software, misreading a weather forecast, or skipping a line on a checklist.

TEM teaches Pilots to constantly look out for threats, anticipate how those threats could lead to an operational error, and apply immediate countermeasures [such as designating a ground visual observer to handle bystanders] to keep the Aircraft secure.

### The Sterile Cockpit Rule

Borrowed directly from commercial Airline Flight decks, the **Sterile Cockpit Rule** dictates that during critical phases of Flight—specifically takeoff, initial climb, final approach, landing, and emergency recovery maneuvers—there must be zero non-essential conversation or distractions.

The Remote Pilot and ground crew must focus 100% of their mental capacity on tracking the Aircraft and monitoring telemetry screens. Casual joking, phone use, or answering questions from clients must wait until the Drone is safely landed, powered down, and its batteries are disconnected.

# Chapter 29: Flight Planning

- Mission Planning
- Route Analysis
- Risk Assessment

# Chapter 29 Flight Planning

A successful Drone deployment is won or lost before the Aircraft ever leaves its case. While basic manual maneuvering is an essential skill, professional enterprise operations require structured Flight planning to ensure safety, legal compliance, and data accuracy.

The **Flight Planning** module integrates Airspace analysis, terrain evaluation, and risk mitigation into a repeatable, multi-step workflow. This systematic preparation protects your crew, your hardware, and the public during complex missions.

## 1. Mission Planning: Defining Scope and Deliverables

Mission planning is the process of translating a client's real-world requirements—such as an agricultural survey, a construction update, or a volumetric stockpile calculation—into specific, executable Flight profiles.

### Technical Parameter Definition

Before establishing a Flight path, a Remote Pilot must calculate the exact Technical constraints required to capture high-quality data:

- **Ground Sampling Distance [GSD]:** In Aerial mapping, GSD represents the physical distance on the ground covered by a single pixel in an image [e.g., 2\text{ cm/pixel}]. A lower GSD provides higher resolution but requires Flying at a lower altitude, which increases overall Flight time.
- **Sensor Selection:** The payload must align with the mission goals. A thermal sensor is required for utility solar inspections, a multispectral camera for crop health analysis, and a high-

resolution mechanical-shutter RGB camera for engineering-grade topographic mapping.

- **Overlap Ratios:** For automated mapping software to stitch separate images into a single 3D model, pictures must overlap significantly. Standard professional missions require a minimum of **70% frontal overlap** [along the Flight line] and **60% side overlap** [between parallel Flight tracks].

## 2. Route Analysis and Terrain Evaluation

Route analysis involves mapping out the exact physical path your Drone will trace through three-dimensional space, ensuring complete structural clearance from the takeoff pad to the final landing approach.

### Topographical and Obstacle Clearance

Standard digital maps do not show local altitude variations or structural hazards. A professional route analysis requires evaluating three distinct layers of elevation data:

```
[ Airspace Ceiling: Max 400 ft AGL ]

     |

     v  _--_ [Local Wind Gusts]

  [ Drone Path ] ---> Clear of Obstacles

     ^

     |  |---| [Tower / Crane Structure]

  [ Ground Level / Elevation Baseline ]
```

- **Digital Elevation Models [DEM]:** You must study the elevation profile of the underlying ground. If you are mapping a hillside and maintain a flat absolute altitude, your Drone will gradually get closer to the terrain, risking a catastrophic ground collision. Advanced Flight planning software uses DEM data to perform **Terrain Awareness Flying**, automatically adjusting the Drone's altitude to match the contour of the hill.

- **Man-Made Obstructions:** Check local databases and conduct physical site scouting to locate non-mapped structural hazards. Pay close attention to construction cranes, cell towers, guy-wires, and overhead high-voltage power lines.

- **The Return-to-Home [RTH] Clear Path:** Your automated safety RTH altitude must be configured *higher* than the tallest known obstacle in the entire area. If the Drone loses signal behind a 150-foot structure, but your RTH is set to its default 100 feet, the Aircraft will Fly straight into the obstacle during its automated return sequence.

## 3. Comprehensive Risk Assessment Framework

A Flight plan is incomplete without an objective risk assessment. Professional Remote Pilots utilize structured frameworks to systematically identify operational vulnerabilities and implement defensive countermeasures.

**The PAVE Risk Categorization Checklist**

To ensure no risk factors are overlooked during pre-Flight planning, Pilots use the **PAVE** framework to analyze the four core elements of a mission:

- **P - Pilot in Command:** Am I legally qualified, current on my training, and physically fit to Fly? [Apply the IMSAFE checklist here].

- **A - Aircraft:** Is the Drone Airworthy? Are the propellers clean, the batteries healthy, the firmware updated, and the payload securely balanced within the Center of Gravity limits?

- **V - Environment:** What are the active weather systems, wind shear gradients, and visibility conditions? Am I Flying in a Green, Yellow, or Red Airspace zone? Are there localized sources of electromagnetic interference nearby?

- **E - External Pressures:** Am I rushing to finish this Flight due to a tight client deadline or a rapidly draining smartphone battery?

External pressure is a primary driver of skipped lines on checklists and poor decision-making.

### Calculating the Final Go/No-Go Threshold

Once all PAVE factors are identified, the scores are applied to a risk matrix to determine the mission's viability:

| Risk Level | Operational Protocol |
|---|---|
| **Low Risk** | Clear to Fly. Proceed using standard operating checklists. |
| **Medium Risk** | Mitigation required. Implement safety controls [e.g., post a visual observer, adjust the Flight path, or delay takeoff for better weather] before proceeding. |
| **High Risk** | **Strict No-Go.** The mission must be aborted or rescheduled. The probability or severity of an accident is too high to justify the operation. |

By treating Flight planning as a strict, data-driven system, you ensure that every commercial deployment minimizes liability, protects public safety, and guarantees operational success.

# PART VII

# OPERATIONS & ADVANCED FLYING

# CHAPTER 30: PRE-FLIGHT PROCEDURES

- Checklist
- Equipment Inspection
- Weather Review

# Chapter 30 Pre-Flight Procedures

In commercial and advanced unmanned Aviation, the margin for error is razor-thin. A single overlooked structural hAirline fracture or an unexpected localized microburst can mean the difference between a successful data-gathering mission and a catastrophic hull loss.

Pre-Flight procedures are not suggestions—they are the foundational baseline of Aviation risk management. This guide breaks down the core pillars of **Chapter 30: Pre-Flight Procedures**, focusing on the critical trinity of Drone operations: **The Checklist Philosophy, Equipment Inspection, and Weather Review.**

## 1. The Pre-Flight Checklist: Your Operational Safeguard

A professional pre-Flight checklist is not a memory aid; it is a rigid process control tool designed to eliminate human error and confirmation bias. Even the most seasoned Remote Pilots can experience cognitive tunneling under field pressures. Relying purely on memory is a leading root cause of avoidable Aviation incidents.

### The "Read-and-Do" vs. "Challenge-and-Response"

Depending on your crew size, your execution strategy will vary:

- **Single- Pilot Operations [Read-and-Do]:** The Remote Pilot in command [RPIC] reads the line item out loud, executes or verifies the state, and manually checks it off.
- **Crewed Operations [Challenge-and-Response]:** A Visual Observer [VO] or co- Pilot reads the "challenge" [e.g., *"Propellers?"*], and the RPIC physically checks the hardware and answers with the "response" [e.g., *"Secure, unmarked, matched."*].

**Aviation Rule:** Never skip an item with the mindset of "I just checked that five minutes ago." Conditions change, and components shift during transport. If you are interrupted during a checklist, drop back **three steps** to ensure continuity.

## 2. Comprehensive Equipment Inspection

Your Unmanned Aircraft System [UAS] must be physically and digitally audited prior to every single launch. Walk the Aircraft in a consistent, clockwise direction every time to ensure your visual sweep covers 100% of the hardware.

### Physical Airframe and Propulsion

- **Structure & Arms:** Inspect all carbon fiber components, booms, and landing gear for stress fractures, delamination, or loose folding-mechanism brackets.
- **Motors:** Spin each rotor manually by hand. Listen closely for grinding or resistance, which indicates dust infiltration or a failing bearing. Verify motor mounts are firmly torqued.
- **Propellers:** Gently flex each blade from root to tip. Check for deep nicks, leading-edge chips, or warping. Ensure that clockwise [CW] and counter-clockwise [CCW] markings precisely match their designated motor assignments, and that quick-release or screw-down locks are completely seated.

### Power & Battery Safety

- **Physical Integrity:** Inspect the lithium polymer [LiPo] casings for any signs of **swelling or bloating**, punctures, or deep dents. A swollen battery is an immediate Flight hazard and must be safely retired.
- **Contacts:** Check that gold-plated battery terminals are clean, free of carbon scoring, and show no signs of arc damage.
- **Secure Seating:** Insert the pack and listen for the physical "click" of the locking tabs. Pull back firmly on the pack to guarantee it cannot back out under high-G maneuvers or vibrations.

**Avionics, Link, and Payload**

- **Sensors & Lens:** Clean all internal vision positioning sensors, obstacles-avoidance cameras, and your primary payload lens with a microfiber cloth.
- **Gimbal Lock:** Always remove the mechanical gimbal clamp before powering on the Aircraft to prevent burning out the stabilization pan/tilt motors.
- **Data & Firmware:** Confirm that your Ground Control Station [GCS] and Aircraft are on stable, matching firmware versions. Ensure the onboard high-speed SD card is formatted and contains enough capacity for the planned mission.

## 3. Pre-Flight Weather Review

A micro-UAV reacts dramatically to micro-climates. A weather forecast that is perfectly fine for a manned commercial Airliner can be completely un-Flyable for an industrial Drone.

Before committing to a Flight, review these critical environmental parameters against your specific Aircraft's operational limitations

- **Wind Speed and Gust Factor:** Compare current ground readings and modeled upper-Air wind charts at your planned operating altitude. A common error is ignoring the *gust factor*—if sustained winds are 15 knots but gusts reach 28 knots, your Aircraft's motor authority may be overridden during a windward turn.
- **Visibility and Cloud Ceiling:** Under most global standard regulations [such as FAA Part 107 or EASA Open/Specific rules], you must maintain a minimum of **3 statute miles** of visibility. Furthermore, your Flight path must remain at least **500 feet below** and **2,000 feet horizontally** away from any cloud structures.
- **Temperature and Dew Point Spread:** Pay close attention to the temperature/dew point spread. When the temperature drops within 2–3°C of the dew point, the Air reaches saturation, creating a high risk for sudden fog formation or lens condensation.

- **Air Density & Thermal Efficiency:** High ambient temperatures significantly reduce Air density [high density altitude]. This requires your rotors to spin faster to create the same amount of lift, causing rapid battery drain and heavily reducing overall Flight time.

## 4. The Unified Pre-Flight Walkthrough

The following step-by-step sequence represents the operational workflow that occurs once you arrive on-site, connecting environmental assessment directly to system startup.

**1.Site & Airspace Verification:**Upon Field Arrival.

Confirm local Airspace authorizations [e.g., LAANC digital approval]. Walk the site boundary to identify overhead lines, localized radio frequency [RF] interference hazards, and establish a clear 25-foot safety radius for your Takeoff and Landing Zone [TLZ].

**2.Environmental Audit:**10 Minutes Before Launch.

Measure real-time surface wind speeds with a handheld anemometer. Verify localized Aviation weather reports [METARs] to ensure no incoming active precipitation or sudden drop in cloud ceilings.

**3.Physical Pre-Flight Inspection:**Cold Aircraft State.

Execute a full structural audit. Check Airframe integrity, verify propeller locking mechanisms are secure, and remove the gimbal clamp. Insert and lock the Flight battery.

**4.Ground Control Initialization:**System Power-On Sequence.

Power on your Remote controller / Ground Control Station app **first**. Set your failsafe behavior parameters [e.g., Return-to-Home [RTH] altitude set safely above the highest local obstacle].

**5.Aircraft Boot & Avionics Check:**Powering the Drone.

Power on the Aircraft. Allow the internal guidance systems to complete their self-test. Verify that the GPS/RTK networks have achieved a stable satellite 3D lock, and ensure the Home Point is accurately updated on your map view.

### 🛑 The Final Go/No-Go Decision

The pre-Flight process ends with a definitive binary decision made by the Pilot in Command. If any single line item on your checklist fails, if a warning light persists, or if a sudden weather shift occurs, you must call a **No-Go**. True professionalism in Aviation is defined by the discipline to keep the Aircraft on the ground when the data counsels caution.

# CHAPTER 31: STANDARD FLIGHT OPERATIONS

- **Launch Procedures**
- **In-Flight Monitoring**
- **Landing Procedures**

# Chapter 31 Standard Flight Operations — The Professional Drone Pilot's Field Guide

A successful commercial Drone deployment is defined by structured precision. Once your pre-Flight checks are complete, you transition into the active Flight window. This phase requires tight operational control, continuous data analysis, and complete situational awareness.

This article details the absolute Flight standards for **Chapter 31: Standard Flight Operations**, breaking down the exact mechanics of **Precision Launch Procedures, Active In-Flight Monitoring, and Controlled Landing Protocols.**

## 1. Precision Launch Procedures: Overcoming Ground Turbulence

The launch phase is a dynamic transition where a Drone changes from a stationary ground object into a stabilized Aircraft. This phase requires clean execution to safely clear localized aerodynamic hazards.

[ Stable Initial Hover: 6–10 ft AGL ] <--- Verify Telemetry & Flight Control Stability

^

| [Smooth, Continuous Throttle Ascent]

[ Ground Effect Zone: 0–3 ft ] <--- High Turbulence / Avoid Hovering Here

^

| [Combination Stick Command [CSC] / Motor Idle]

[ Launch Pad / Ground State ]

### Motor Arming & The Idle Check

Before taking off, the Remote Pilot in Command [RPIC] must formally arm the Aircraft's Flight controller. On most commercial and enterprise Drone platforms, this is accomplished via a **Combination Stick Command [CSC]**—typically pulling both control sticks down-and-inward or down-and-outward simultaneously.

Once armed, the brushless motors engage at a low idle speed. The Pilot must pause for 2 to 3 seconds to execute a visual and auditory audit: verify that all rotors are spinning uniformly, and listen for any abnormal high-frequency whistling or rattling that indicates an overlooked mechanical issue.

### Navigating the Ground Effect Zone

When a multi-rotor Drone sits near a flat surface, its high-velocity propeller downwash hits the ground and bounces back up. This creates a highly turbulent, unpredictable cushion of Air known as **ground effect**. This zone generally extends from the surface up to an altitude equal to roughly the diameter of the Drone's rotor span [typically the first 3 feet of Flight].

To launch safely:

- Apply a smooth, continuous upward input on the throttle stick to lift the Aircraft cleanly *through* the ground effect zone.
- Do not pause or attempt to hover within 1 or 2 feet of the ground, as the recycled Air currents will cause the Aircraft to skitter and drift erratically.
- Climb directly to a baseline holding altitude between **6 and 10 feet Above Ground Level** [AGL]. Hold this position for 10 to 15 seconds. This pause allows you to verify that the Aircraft has locked its GPS coordinates perfectly, its altitude-hold is stable, and it responds accurately to minor stick inputs before you command it higher.

## 2. In-Flight Monitoring: The Telemetry Cross-Check

Once your Drone climbs to its designated mission altitude, your primary responsibility shifts to continuous systems monitoring. A professional Pilot never stares blankly at the live camera feed; instead, they maintain a systematic cross-check between the physical Aircraft in the sky and the data processing on their Ground Control Station [GCS].

**The Pilot Visual Scan Cycle**

To maintain true Visual Line of Sight [VLOS] and total situational awareness, implement a structured visual scanning loop. Never lock your eyes onto your mobile display or tablet for more than 3 to 5 seconds at a time.

**Critical Telemetry Metrics to Track**

- **Absolute Voltage [V] vs. Battery Percentage:** Do not rely solely on the simple percentage graphic icon. Monitor the absolute voltage of the pack and look for individual cell balance readouts. If a single cell's voltage drops significantly lower than the others under load, your battery is at risk of cell failure—command an immediate landing.
- **Signal Strength Indicators [RSSI]:** Track both your **Uplink Control Link [RC]** and your **Downlink Video/Data Link**. If either signal drops below a safe operational margin [typically 40%], pause your mission, adjust your ground station antenna alignment, or bring the Aircraft closer to prevent an automatic Return-to-Home [RTH] trigger.
- **Groundspeed vs. Airspeed:** If your Drone is programmed to Fly an automated mapping grid at 20 km/h, but its groundspeed drops down to 5 km/h on a specific leg, it is battling a high headwind. This will drain your battery cells significantly faster than your software's original Flight plan calculated.

## 3. Controlled Landing Procedures: Securing the Touchdown

A Flight is never truly successful until the Aircraft is safely anchored on the landing pad with its motors completely powered down. The landing phase demands high focus because the Drone must re-enter turbulent Air currents close to ground-level obstructions.

1.**Approach and Alignment:**Altitude: 10–15 ft AGL.

Navigate the Aircraft back to your designated landing zone. Bring it to a stable hover directly over the center of your physical landing pad. Ensure the Drone's nose is pointed completely away from you [Tail-In orientation] so your manual control stick responses remain fully intuitive.

2.**Clearance Verification:**Radius: 3 Meters.

Perform a rapid visual scan of the landing footprint. Verify that no clients, bystanders, or animals have wandered into your 3-meter safety radius while your eyes were tracking the final approach.

3.**Controlled Descent:**Rate: Under 1 m/s.

Apply a gentle, steady downward pressure on your throttle stick. Maintain a slow, vertical descent path. Avoid making sudden, aggressive corrections near the surface to prevent entering a **Vortex Ring State**—an aerodynamic condition where a Drone sinks into its own turbulent downwash, destroying its ability to generate lift.

4.**Firm Contact and Disarming:**Touchdown State.

Guide the Aircraft down until its landing gear makes firm, level contact with the ground pad. Once down, hold the throttle stick down completely for 3 continuous seconds, or execute your system's manual disarm command. This action tells the Flight computer to cut all voltage to the Electronic Speed Controllers [ESCs], stopping the propellers instantly.

5.**System Power-Down Sequence:**Post-Flight Protocol.

Once the rotors stop spinning, approach the Aircraft. Always turn off the **Aircraft battery first**, then power down your **Remote Controller**. This ensures the Drone is never left powered on in an active state without an active, authoritative control signal source nearby.

## 4. Operational Summary Matrix

Use this quick-reference table to memorize your key responsibilities during each critical Flight phase:

| Operational Phase | Primary Objective | Critical Risk | Key Mitigation Strategy |
|---|---|---|---|
| **1. Launch Procedures** | Clean transition to clear, undisturbed Airspace. | Ground-effect turbulence and skittering. | Apply smooth, continuous throttle to a stable 6–10 ft AGL hover before moving. |
| **2. In-Flight Monitoring** | Consistent data capture and systems tracking. | Telemetry blind spots and unexpected link loss. | Maintain a rigid, repetitive visual scan cycle between the sky and your GCS screen. |
| **3. Landing Procedures** | Secure touchdown and complete motor disarming. | Vortex Ring State [VRS] and Aircraft tip-overs. | Maintain a slow descent rate [< 1 m/s] in a tail-in orientation over a cleared landing pad. |

### 📝 The Pilot's Logbook Requirement

The moment your standard Flight operation terminates with a successful power-down, your legal recording obligations begin. Always log your total Flight time, battery cycle numbers, structural component status, and any minor telemetry anomalies immediately into your physical or digital Pilot logbook to maintain complete fleet Airworthiness records.

Standard Flight Operations

- Launch Procedures
- In-Flight Monitoring
- Landing Procedures

# Chapter 32: Autonomous Flight Operations

- **Mission Planning**
- **Automated Routes**
- **Waypoint Flying**

# Chapter 32 Autonomous Flight Operations

Manual Flight control is an essential skill for every Remote Pilot, but the true scalability of commercial Drone Technology relies on automation. **Autonomous Flight Operations** use advanced Flight planning software, satellite navigation networks, and onboard telemetry processors to execute missions with high mathematical precision.

By replacing manual stick inputs with software-driven commands, operators can perform complex Aerial mapping, systematic structural inspections, and long-range linear monitoring with total repeatability.

## 1. Automated Mission Planning Fundamentals

Autonomous Flight operations do not mean clicking a button and letting the Drone figure out the mission. True automation requires a comprehensive, data-driven planning phase on a Ground Control Station [GCS] before the Aircraft ever initializes its launch sequence.

### Defining Spatial Boundaries and Safety Buffers

The first step in autonomous planning is establishing a digital boundary—frequently using a polygon tool overlaid onto high-resolution satellite basemaps.

- **The Geofence Anchor:** The Pilot must map out a strict operational footprint. The software builds a virtual wall [geofence] around this boundary. If a software glitch or wind shear pushes the Drone outside this grid, the system automatically triggers an abort or an immediate landing sequence.
- **Safety Overlap and Clearance Zones:** Automated Flight lines must be placed with a safety clearance buffer [typically at least 30 to 50 feet] away from known manual obstacles, such as cell towers

or construction cranes, to account for standard GPS positioning tolerances.

## 2. Automated Route Structures and Enterprise Profiles

Different industries require different automated Flight path geometries. Understanding the "shape" of your data collection path is critical to optimizing battery life and image quality.

### Common Flight Geometry Formats

#### 1. The Lawnmower Grid Matrix [Aerial Mapping & Photogrammetry]

Used extensively for topographical surveys, agricultural analysis, and construction progress tracking. The Drone flies a highly structured back-and-forth grid pattern over a specified area. The GCS calculates the Drone's Flight speed and camera capture intervals based on two primary variables:

- **Frontal Overlap:** The percentage of image overlap along a single Flight line [standard is **70% to 80%**].
- **Side Overlap:** The percentage of image overlap between parallel Flight tracks [standard is **60% to 70%**].

#### 2. Corridor Mapping [Linear Asset Inspection]

Designed specifically for tracking continuous, narrow structures such as pipelines, electrical power transmission lines, railway tracks, and highways. The software generates a streamlined, single-line or narrow multi-track path that mirrors the asset's curvature, maximizing coverage while keeping the Drone well away from structural components.

#### 3. Point of Interest [POI] Orbit [Vertical Inspection]

Used for scanning concentrated vertical assets like telecommunication towers, wind turbines, smokestacks, or storage tanks. The Pilot defines the center coordinate of the asset, and the autonomous system commands the Drone to execute a perfect geometric circle around the structure, keeping the payload camera locked inward toward the asset face at a constant distance.

## 3. Waypoint Flying and Parameter Configuration

Waypoint Flying is the core building block of all Drone automation. A waypoint is a specific three-dimensional point in space defined by latitude, longitude, and relative altitude coordinates [X, Y, Z]. A complete autonomous route is simply a sequential string of these waypoints linked together by a digital Flight computer.

**Micro-Managing Individual Waypoint Actions**

Modern enterprise waypoint software allows Pilots to assign highly specific, localized actions to the Drone at *each* distinct node along its path. Rather than just Flying past a point, the Drone can be programmed to pause and execute a variety of tasks:

## 4. The Autonomous Mission Workflow

Executing an autonomous Flight requires a strict handoff sequence between manual Pilot oversight and automated system execution.

**1.Mission Programming & Validation:**On-Ground State.

Build the waypoint route or grid matrix within the GCS software. Run the system's digital simulator tool to check for altitude errors, terrain clearance issues, and estimated battery depletion parameters.

**2.Data Upload and Handshake:**Pre-Launch Connection.

Power on the GCS and the Drone. Once a stable satellite 3D lock is established, upload the mission file to the Drone's internal Flight computer. The Flight controller cross-checks the parameters and returns a "Ready to Launch" confirmation token.

**3.Manual Takeoff & Systems Validation:**Altitude: 6–10 ft AGL.

Always launch the Drone manually first. Bring it to a stable hover at eye level for 10 seconds. Check the telemetry feed to verify that the home point is locked and that no sensor errors are active before engaging automation.

**4.Mission Engagement & Oversight:**Active Automated Phase.

Flip the designated switch on your Remote controller to execute the autonomous mission. The Drone will automatically climb to its starting waypoint and begin its route. **The Pilot must keep their hands on the control sticks at all times**, tracking the Drone's position visually to be ready to override the automation instantly if an obstacle appears.

**5.Automation Termination & Safe Landing:**Mission Conclusion.

Once the final waypoint is cleared, the Drone will execute its pre-programmed end-of-mission behavior [typically hovering at the final point or initializing an automatic Return-to-Home sequence]. The Pilot can choose to let the system land automatically or take manual control to guide the Aircraft down onto the landing pad.

**⚠ The Manual Override Absolute Rule**

Autonomy does not mean a Pilot is relieved of responsibility. Under all global Aviation regulations, the Remote Pilot in Command [RPIC] retains legal liability for the Flight. If a crewed Aircraft enters the workspace, or if a sensor anomaly causes a Flight path deviation, the Pilot must instantly flip the Flight mode switch out of "Auto" and back into "Manual/GPS" mode to take direct mechanical control of the Aircraft.

# CHAPTER 33: BVLOS OPERATIONS

- **Beyond Visual Line of Sight**
- **Future Regulations**
- **Commercial Applications**

# Chapter 33 BVLOS Operations

Manual Flight within your natural line of sight represents the foundational tier of Drone Piloting. However, the true economic and operational revolution of unmanned Aviation occurs when the Aircraft breaks free from the physical constraints of the Pilot's eyes.

**Beyond Visual Line of Sight [BVLOS]** operations allow Unmanned Aircraft Systems [UAS] to Fly long-distance, cross-country corridors using advanced telemetry networks, Remote sensor arrays, and automated traffic management frameworks.

## 1. Defining Beyond Visual Line of Sight (BVLOS)

In standard Drone operations, the Remote Pilot in Command [RPIC] must maintain continuous, unaided **Visual Line of Sight [VLOS]** with the Aircraft to track its position, monitor its heading, and spot nearby obstacles or crewed Aircraft.

**The Spectrum of Sight Limitations**

- **Visual Line of Sight [VLOS]:** The Aircraft is clearly visible to the Pilot without binoculars or telescopes at all times. This limits real-world operations to a radius of roughly 1 to 1.5 kilometers depending on weather and Drone size.
- **Extended Visual Line of Sight [EVLOS]:** The Pilot cannot see the Drone, but a network of trained **Visual Observers [VOs]** are stationed along the Flight path. These observers maintain a human relay chain, communicating via two-way radios to track the Drone and warn the Pilot of incoming traffic.
- **Beyond Visual Line of Sight [BVLOS]:** The Drone flies entirely out of view of the Pilot and any ground observers. The Pilot relies

100% on instruments, onboard cameras, digital navigation maps, and automated telemetry feeds to maintain situational awareness.

### Technical Prerequisites for BVLOS Flight

To safely operate an Aircraft over several dozen kilometers without human eyes tracking it, the system requires advanced Technical redundancies:

- **Command and Control [C2] Data Links:** Traditional radio frequencies [2.4\text{ GHz} or 5.8\text{ GHz}] lose signal just past a few kilometers. BVLOS operations use robust long-range links, primarily **4G/5G cellular networks** or **Satellite Communications [SatCom]**, allowing the Drone to be controlled from a ground station halfway across the country.
- **Detect and Avoid [DAA] Systems:** Since the Pilot cannot visually spot a low-Flying medical helicopter or light sport Aircraft, the Drone must spot hazards itself. BVLOS platforms carry integrated DAA suites consisting of **Airborne Radar**, **optical camera arrays**, and **ADS-B In receivers** [Automatic Dependent Surveillance-Broadcast] that automatically alter the Drone's path to avoid conflicting traffic.

## 2. High-Value Commercial Applications

Removing the visual line-of-sight barrier changes Drones from localized inspection tools into massive industrial infrastructure networks.

## 3. The Future of Global and Indian Regulations

Because BVLOS operations share the sky directly with traditional crewed Aviation, regulatory bodies are building highly structured, risk-based frameworks to move away from case-by-case waivers toward routine, standardized approvals.

### The Risk-Based Approach [SORA]

The global gold standard for approving BVLOS operations is the **Specific Operations Risk Assessment [SORA]** framework, developed by JARUS and adopted extensively across Europe and Asia. Rather than applying a one-size-fits-all rule, SORA requires operators to analyze their mission across two distinct profiles:

1. **Ground Risk Class [GRC]:** What is the probability and severity of the Drone hitting a person on the ground? [e.g., Flying over a deserted desert vs. Flying over a crowded city street].
2. **Air Risk Class [ARC]:** What is the probability of colliding with another Aircraft? [e.g., operating at 200 feet in Remote Airspace vs. Flying near an Airport approach corridor].

The combination of these classes dictates the strictness of the Technical and operational safeguards the Pilot must demonstrate before receiving a Flight authorization token.

**Digital Sky and UTM Integration in India**

The Directorate General of Civil Aviation [DGCA] and the Ministry of Civil Aviation in India have charted a clear path forward for standardized BVLOS operations via the **Digital Sky Platform** and the rolling implementation of **Unmanned Aircraft System Traffic Management [UTM]**:

- **BVLOS Experimental Frameworks:** India has successfully established multi-agency consortia to test long-range medical deliveries [vaccines, emergency medicine] and automated agricultural mapping within dedicated, restricted test corridors.
- **Digital Sky Integration:** The future regulatory environment integrates the "No Permission, No Takeoff" [NPNT] architecture into real-time UTM networks. As Airspace is digitized, the platform will automatically manage separations between multiple autonomous BVLOS Drones using machine-to-machine APIs.
- **The Drone Corridor Concept:** Regulatory bodies are designing permanent, designated three-dimensional "Drone Corridors"—segregated blocks of low-altitude Airspace reserved exclusively for commercial cargo and data missions, ensuring complete isolation from traditional crewed Flight paths.

## 4. Standard BVLOS Mission Safety Workflow

Operating a BVLOS mission requires a strict, multi-layered risk validation loop that transitions the Pilot from local site awareness to wide-area system monitoring.

**1.Strategic Route Pre-Validation:**GCS Network Setup.

Map out the cross-country waypoint route on your digital GCS. Cross-reference the path with real-time digital elevation models [DEM] to ensure structural ground clearance. Verify that cellular/satellite signal strength profiles remain continuous across the entire Flight track.

**2.Hardware Link Handshake:**Avionics Power-On.

Power on the long-range command station and the Drone. Confirm the activation of dual-redundant C2 links [e.g., primary 5G network with a secondary SatCom backup]. Execute a digital handshake to confirm the onboard Flight computer has securely cached the automated return-to-home parameters.

**3.Sensor and DAA Initializations:**Failsafe Verification.

Verify that the onboard Airborne radar, optical camera vision arrays, and ADS-B In receiver are active and processing live data strings cleanly. Ensure the local weather METAR maps confirm zero icing conditions along the distant Flight route.

**4.The Local Launch Phase:**Altitude: 0–100 ft AGL.

Execute a manual or automated vertical takeoff. Keep the Drone within your local Visual Line of Sight [VLOS] for the first 60 seconds of Flight, performing a minor system holding pattern to verify correct propulsion output and stable sensor tracking before sending it out on its route.

**5.Transition to Instrument Oversight:**Active En-Route State.

Command the Drone to enter its automated route profile as it leaves your line of sight. The Pilot's role transitions to an **Airspace manager**—monitoring the live telemetry dashboard, tracking individual cell health voltages, and keeping ready to initiate a Remote override command if an automated traffic warning flashes.

### The Redundant Failsafe Absolute Rule

Every certified commercial BVLOS platform must be equipped with an un-jammable, completely independent **Termination System [FTS]**. If the primary Flight computer experiences a total software freeze and stops responding while the Drone is Flying out of sight, the Pilot must have a separate, dedicated radio switch to cut all electrical power to the motors instantly or deploy an emergency ballistic parachute to prevent a long-range Flyaway event.

# CHAPTER 34: NIGHT OPERATIONS

- **Equipment Requirements**
- **Safety Measures**

# Chapter 34 Night Operations

Operating a Drone after dark changes the risk profile of a mission completely. When daylight drops away, a Pilot's natural depth perception, structural awareness, and ability to spot hazards drop with it.

**Night Operations** require specialized hardware setups, strict regulatory compliance, and unique physiological adaptations. Mastering these safeguards allows commercial Pilots to tap into critical low-light workflows, including industrial thermal imaging, nighttime security surveillance, and midnight agricultural spraying.

## 1. Defining "Night" and Regulatory Frameworks

A common mistake among novice Pilots is assuming night rules only apply when the sky is completely pitch-black. Aviation regulatory bodies define low-light periods with strict mathematical precision.

**The Twilight Window**

- **Evening Civil Twilight:** The period from official sunset to **30 minutes after sunset**.
- **Morning Civil Twilight:** The period from **30 minutes before official sunrise** to official sunrise.
- **Night Operations:** Any Flight occurring *after* evening civil twilight ends or

**Global Regulatory Requirements**

Historically, Flying at night required complex individual waivers. Today, modern regulations have standardized night Flying by embedding it directly into commercial Pilot qualifications, provided specific equipment conditions are satisfied:

- **FAA Part 107 [United States]:** Commercial Pilots can operate at night without a specific waiver, provided they have completed updated low-light training and the Aircraft is equipped with compliant anti-collision lighting.
- **DGCA Drone Rules [India]:** Night Flying is restricted by default. To operate between sunset and sunrise, commercial Pilots must hold a valid Remote Pilot Certificate [RPC], file a detailed Flight plan, and secure explicit digital approval through the **Digital Sky / eGCA portal** [typically requiring operations within authorized corridors or specific low-risk zones].

## 2. Mandatory Equipment Requirements

You cannot Fly a stock enterprise Drone at night straight out of the box. Consumer and industrial Aircraft feature factory LEDs, but these are classified as **positional/navigation lights** [used to determine the Drone's heading at short distances] and do not meet legal safety minimums for night operations.

### Anti-Collision Lighting Criteria

To operate at night safely, the Aircraft must carry an auxiliary, high-intensity **Anti-Collision Strobe Light**. This hardware must meet three strict Technical specifications:

1. **3-Statute-Mile Visibility:** The strobe must be bright enough to be spotted by a crewed Aircraft Pilot from a distance of at least 3 statute miles [approximately 4.8\text{ kilometers}] in clear atmospheric conditions.
2. **Flash Rate Threshold:** The light cannot remain solid or pulse slowly. It must strobe at a rate between **40 and 100 flashes per minute** to instantly distinguish it from ground-level streetlights or stars.
3. **360-Degree Hemispherical Coverage:** The strobe is ideally mounted to the **top** of the Aircraft's fuselage. Because manned

Aircraft Fly above Drones, a bottom-mounted light is completely blocked by the Drone's own body, rendering it useless to an oncoming helicopter or plane.

## 3. Human Physiology: The Illusion of Sight

Flying at night requires understanding how the human eye processes light. The biological adaptation to darkness introduces critical physiological vulnerabilities that a Pilot must actively mitigate.

# Chapter 35: Emergency Operations

- Loss of GPS
- Loss of Signal
- Battery Failure
- Crash Management

# Chapter 35 Emergency Operations

No matter how meticulous your pre-Flight planning is, real-world Aviation presents unpredictable variables. When a component fails, the transition from a standard Flight to an emergency scenario happens in milliseconds.

Managing an inFlight crisis requires a combination of programmed automated fail-safes and instant, deliberate manual intervention. A Pilot must maintain complete emotional composure, quickly diagnose the system failure, and execute structured emergency recovery checklists.

## 1. Loss of GPS [GPS Attenuation / Flyaway]

When an Unmanned Aircraft System [UAS] loses connection with its satellite constellations [GPS/GLONASS/NavIC], it loses its ability to calculate its precise spatial position, groundspeed, and hover stability.

[ GNSS Signal Fail

**The Immediate Symptom**

The Drone will automatically drop out of **Position Mode [P-Mode]** and enter **Attitude Mode [ATTI Mode]**. In ATTI mode, the internal barometers still maintain the Drone's altitude [height], but the automatic horizontal braking is completely disabled. If the Pilot releases the control sticks, the Drone will not stop; it will drift rapidly with the prevailing wind velocity.

**Step-by-Step Pilot Recovery Loop**

**1.Acknowledge Mode Change:**Telemetry Assessment.

Identify the visual or audible "GPS Signal Loss" or "ATTI Mode Active" warning on your Ground Control Station [GCS]. Instantly check your screen's satellite count indicator to verify total GNSS attenuation.

**2.Establish Active Manual Attitude Control:**Stick Input Shift.

Do not let go of the control sticks. Treat the Aircraft as a purely manual aerodynamic platform. Apply continuous, active stick corrections to counter any wind drift and bring the Drone to a stable, visual physical hover.

**3.Inhibit Automated Return-to-Home:**RTH Prevention.

**DO NOT press the automated Return-to-Home [RTH] button.** Because the system lacks valid GPS coordinates, triggering RTH could cause the Drone to calculate an erroneous home path, resulting in an un-controlled Flyaway.

**4.Navigate Home via Visual Reference:**VLOS Navigation.

Turn the Drone's camera toward your location or look directly at its physical orientation lights. Fly the Aircraft backward toward your designated landing pad using strict visual line-of-sight tracking and instrument cross-references.

## 2. Loss of Signal [Lost Link]

A **Lost Link** scenario occurs when the communication link between the Ground Control Station transmitter and the Drone's onboard receiver is broken. This can be caused by radio frequency interference, physical terrain obstructions [shading], or Flying past maximum range limits.

### Uplink vs. Downlink Failure

- **Downlink Failure:** The video transmission or telemetry feed cuts out, but the stick inputs still affect the Drone. The Pilot can visually navigate the Drone home safely if it remains within physical line of sight.
- **Uplink Failure:** The control link itself breaks. The Drone ceases to receive inputs from the Pilot's sticks and must rely entirely on its pre-programmed software behaviors.

### Programmable Failsafe Implementations

Modern enterprise systems allow the Remote Pilot to define precise automated behaviors when a lost link condition persists past a specific temporal threshold [typically 3 to 5 seconds]:

- **Return-to-Home [RTH]:** The default professional setting. The Drone automatically climbs to a pre-configured clear altitude, flies a straight vector back to the coordinates of the launch pad, and executes an automated landing.
- **Hover in Place:** The Drone stops moving and maintains its exact altitude and position. This is ideal when operating inside dense tree canopies or beneath structural overhangs where an automatic climb would result in an immediate overhead strike.
- **Land Immediately:** The Drone terminates its horizontal track and begins a steady vertical descent to the ground from its current position. This profile is used when Flying within highly dynamic, complex Airspaces where long-range autonomous transit introduces excessive risk.

## 3. Critical Battery Failure

Battery emergencies stem from manufacturing cell degradation, operating in extreme cold, or miscalculating wind resistance on the outbound Flight leg, which leaves insufficient reserves for the return trip.

### Voltage Sag vs. Thermal Runaway

- **Voltage Sag:** Under heavy throttle demands or high headwind resistance, a battery's voltage can drop sharply. While the overall capacity might indicate 30%, the instantaneous voltage drops below the critical threshold required to sustain motor output, triggering a sudden emergency descent.
- **Thermal Runaway:** If internal chemical cells crack, short, or exceed critical thermal ceilings [65^\circ\text{C}+], the battery enters an uncontrollable self-heating loop. This can lead to a severe

catastrophic mid-Air electrical fire and immediate propulsion failure.

**Safe Capacity Allocation**

To protect hardware assets and ensure public safety, professional Pilots structure their Flight schedules around a **Three-Tier Power Management Model**.

## 4. Post-Crash Management Protocols

If a collision or system failure forces the Aircraft down, the Pilot's responsibilities pivot instantly from Flight management to crisis control and regulatory compliance.

**Immediate Ground Safety Prerequisite**

If the Drone crashes into live infrastructure [such as an electrical substation or high-voltage power lines], **DO NOT attempt to retrieve the hardware yourself.** Cordon off the area to keep bystanders away and contact local emergency services immediately.

**Incident Management Checklist**

- **Step 1: Power Isolation:** If the Aircraft is safely accessible, immediately approach the debris and extract the Flight battery. This isolates the power grid and prevents a potential lithium-polymer thermal fire.
- **Step 2: Log Export and Storage:** Before cycling power to your Remote controller, download the active Flight telemetry logs [.DAT or .TXT files]. These files preserve critical sensor data, motor outputs, and error state codes required to diagnose the root cause of the crash.
- **Step 3: Secure the Perimeter:** Collect all scattered components and debris. Ensure no hazardous chemicals or material fragments are left behind on public or private property.

**Global and Indian Regulatory Reporting Rules**

Accident reporting is an absolute legal requirement across global Aviation frameworks. Under the **DGCA Drone Rules 2021** [and active eGCA/

Digital Sky frameworks in India], standard reporting protocols must be followed:

**The 24-Hour Mandatory Reporting Window:** If a registered Unmanned Aircraft System causes significant property damage or personal injury to any third party, or goes missing entirely, the Remote Pilot in Command **must submit a formal incident report through the Digital Sky Platform within 24 Hours** of the event. Failure to report an accident is a punishable legal offense and can result in the immediate suspension of your Remote Pilot Certificate [RPC].

# Disclaimer

The information, regulations, policies, procedures, market data, technology trends, and industry insights presented in this book are based on sources believed to be accurate and reliable as of **1 June 2026**.

The Global Drone industry, Aviation regulations, government policies, certification requirements, and technological developments are subject to frequent updates and changes. Readers are strongly advised to verify the latest information, notifications, circulars, guidelines, and regulatory requirements from the relevant authorities, including the Directorate General of Civil Aviation (DGCA), Government of India, and other applicable organizations before making any operational, educational, financial, or career-related decisions.

While every effort has been made to ensure the accuracy and completeness of the information contained in this book, the author and publisher make no warranties or representations regarding its accuracy, completeness, or suitability for any specific purpose. The author and publisher shall not be held liable for any loss, damage, inconvenience, or consequences arising from the use of the information provided herein.

This book is intended solely for educational, informational, and career guidance purposes and should not be considered legal, regulatory, financial, or professional advice.

**All information contained in this book was current and believed to be accurate as of 1 June 2026, prior to publication.**

Asiatic International Corp
Drone.asiaticincorp.org
TRAIN. FLY. SUCCEED.
YOUR SKY. YOUR FUTURE.
BUILD YOUR CAREER IN
DRONE
AVIATION
FLY • LEARN • LEAD
PROFESSIONAL TRAINING
DGCA Certified Courses
by Expert Trainers
PRACTICAL EXPOSURE
Hands-on Flying Experience
& Real World Missions
CAREER OPPORTUNITIES
High Demand Jobs in India
& Around the World
AGRICULTURE
SURVEY & MAPPING
INSPECTION
MEDIA & CINEMATOGRAPHY
DEFENSE & SECURITY
DELIVERY & LOGISTICS
TAKE CONTROL
OF YOUR FUTURE
JOIN NOW >
Drone.asiaticincorp.org
SKY IS NOT THE LIMIT,
IT'S YOUR STARTING POINT.

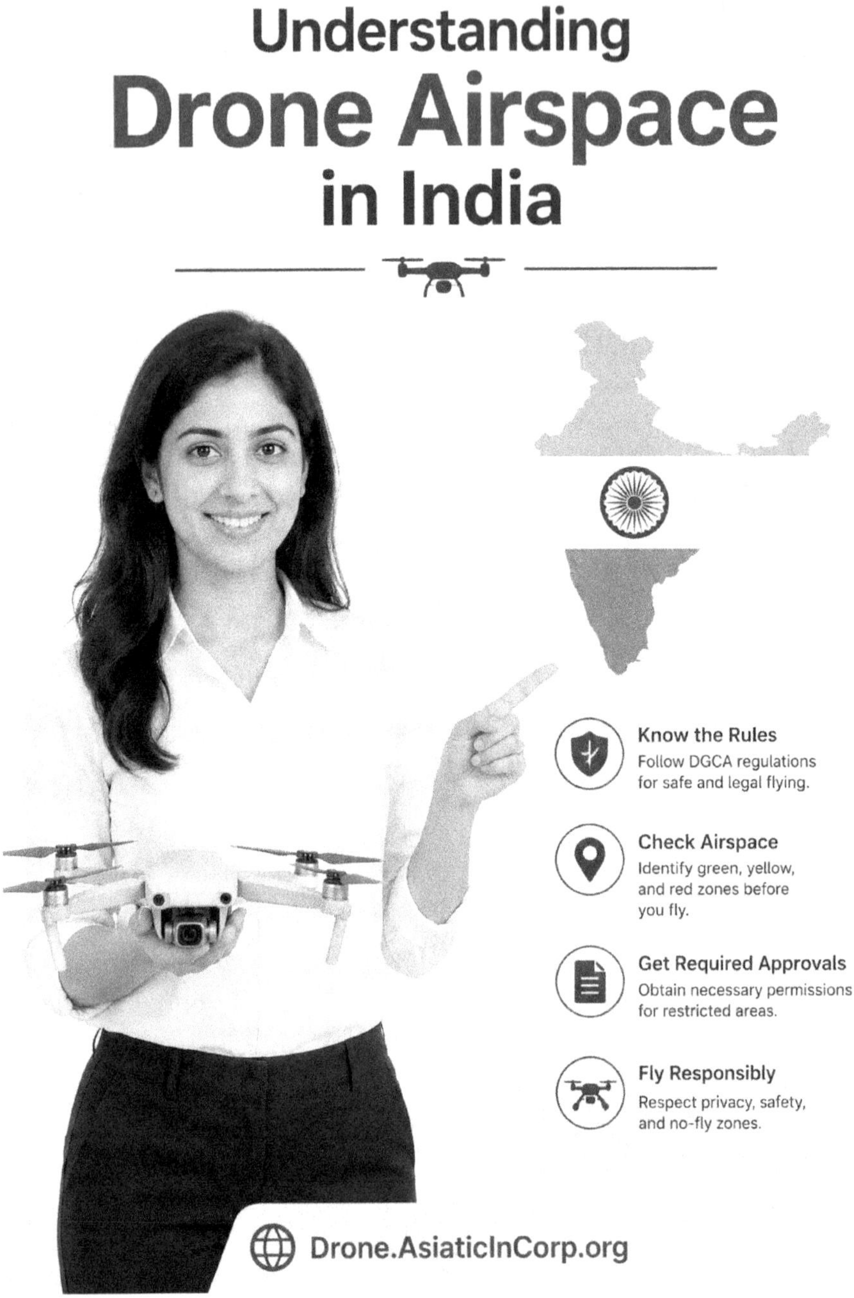
Understanding
Drone Airspace
in India
Know the Rules
Follow DGCA regulations for safe and legal flying.
Check Airspace
Identify green, yellow, and red zones before you fly.
Get Required Approvals
Obtain necessary permissions for restricted areas.
Fly Responsibly
Respect privacy, safety, and no-fly zones.
Drone.AsiaticInCorp.org

Asiatic International Corp
Asiatic
International Corp
(Formerly AirCrews Aviation Pvt. Ltd.)
CONFUSED
AFTER 12TH?
YOU'RE NOT ALONE.
Get the RIGHT CAREER GUIDANCE
before you take the next step.
Capt. Shekhar Gupta
Aviator | Author | Career Guide
Helping Students & Professionals

# Understanding Airspace

## Green Zone

- Low risk. No permission required for nano drones.
- Fly freely within permitted limitations.

## Yellow Zone

- Moderate risk. Permission required before flying.
- Check airspace and obtain approval on Digital Sky.

## Red Zone

- High risk. Flying drones is not allowed.
- Includes airports, military areas, and sensitive locations.

## Controlled Airspace

- Airspace that requires ATC clearance for flights.
- Includes Class C, D, E & G airspace.

## Temporary Restrictions

- Restrictions imposed for a specific time period.
- May apply due to events, VIP visits, disasters, or security reasons.

Drone.AsiaticInCorp.org

BOOKS THAT
INSPIRE, EDUCATE & EMPOWER
BY CAPT. SHEKHAR GUPTA AND CO-AUTHORS
FLY HIGH
A Guide to a Career in Aviation Industry
Capt Shekhar Gupta
From Dreamer To CEO
CAPT. SHEKHAR GUPTA
PILOT'S CAREER GUIDE
Step by Step Learn How to Become an International Airline Pilot
Capt Shekhar Gupta
Nirhba Khajanchi
LEARN TO FLY HERE
YOUR PILOT CAREER GUIDE
CAPT. SHEKHAR GUPTA
LESSONS FOR BUSINESS LEADERS
Navigating the Path to Success: All Essential Lessons for Business Leaders
CAPT. SHEKHAR GUPTA
INTERNATIONAL AIR HOSTESS CAREER GUIDE
YOUR AVIATION CAREER GUIDE
CAPT SHEKHAR GUPTA
SARITA LALWANI
BUILDING HR EXCELLENCE
A Startup Founder's Guide
CAPT. SHEKHAR GUPTA
SHRISHTY SHARMA
MOST SUCCESSFUL FEMALE ENTREPRENEURS OF INDIA
GRISHMA VIJAY
CAPT. SHEKHAR GUPTA
NOW AVAILABLE ON
amazon
AlfaBook.AsiaticInCorp.org
+91 9977513452

भारतीय ड्रोन प्रौद्योगिकी संस्थान
Indian Institute Of Drone Technology
DGCA-Approved
5 Days Remote
Pilot Training
Program
Your Drone Career Starts Here!
The Indian Institute of Drone Technology (IIDT) is excited to announce that our drone industry partner — Prakhar Software Solutions Limited — is now offering a DGCA-approved RPTO training at IIDT Dodi Campus.
WHY CHOOSE THIS PROGRAM?
Fast-track DGCA-approved pilot training
Real flying experience, not just theory
Learn from certified instructors at an RPTO-authorized centre
Designed for beginners as well as professionals
WHAT YOU WILL RECEIVE
1 Study Material
2 3 Hours of Simulator Training
3 4.5 Hours of Flying Practice
4 Official Pilot Logbook
5 Career Guidance & Counseling
6 Certificates:
• DGCA Certificate of Completion
• RPTO Certificate
7 Added benefits:
• 3-Day IIDT Drone Workshop Certificate (at no extra cost)
• IIDT goodies (T-shirt, diaries & more)
PROGRAM DETAILS
Duration: 5 Days
Mode: On-site (Residential)
Venue: IIDT Dodi Campus
Food & Accommodation: Included
COURSE FEE:
₹30,000/-
SPECIAL LAUNCH OFFER:
₹25,000/-
(LIMITED SEATS)
CALL / WHATSAPP
+91 9977513452
Indian Institute of Drone Technology, Dodi (MP)

Pan-India Tourism Network Directory

**Dir.TourTravelNet.org**

DGCA-Approved
5 Days Remote
Pilot Training
Program
Your Drone Career Starts Here!
The Indian Institute of Drone Technology (IIDT) is excited to announce that our drone industry partner — Prakhar Software Solutions Limited — is now offering a DGCA-approved RPTO training at IIDT Dodi Campus.
WHY CHOOSE THIS PROGRAM?
Fast-track DGCA-approved pilot training
Real flying experience, not just theory
Learn from certified instructors at an RPTO authorized centre
Designed for beginners as well as professionals
WHAT YOU WILL RECEIVE
Study Material
3 Hours of Simulator Training
4.5 Hours of Flying Practice
Official Pilot Logbook
Career Guidance & Counseling
Certificates
DGCA Certificate of Completion
RPTO Certificate
Added benefits
3-Day IIDT Drone Workshop Certificate (at no extra cost)
IIDT goodies (T-shirt, diaries & more)
PROGRAM DETAILS
Duration: 5 Days
Mode: On-site (Residential)
Venue: IIDT Dodi Campus
Food & Accommodation: Included
COURSE FEE: ₹30,000/-
SPECIAL LAUNCH OFFER! ₹25,000/-
(LIMITED SEATS)
CALL / WHATSAPP +91 9977513452
Indian Institute of Drone Technology, Dodi (MP)
Email: info.rpto@prakharsoftwares.com
Web: Drone.AsiaticInCorp.org
GET CERTIFIED. GAIN SKILLS.
FLY HIGH. BUILD YOUR FUTURE.
REMOTE PILOT CAREER GUIDE
CAPT SHEKHAR GUPTA
REMOTE
PILOT
CAREER GUIDE
FROM BEGINNER TO
PROFESSIONAL DRONE CAPTAIN
STEP BY STEP GUIDE
LEARN The Fundamentals
BUILD Your Skills
EARN Certifications
LAUNCH Your Career
WORK Anywhere in the World
DRONE CAPTAIN
AUTHOR:
CAPT SHEKHAR GUPTA
AIRPLANE AND REMOTE PILOT

www.ingramcontent.com/pod-product-compliance
Lightning Source LLC
LaVergne TN
LVHW052044200926
848481LV00005B/1730